A TREATISE

ON THE

RESISTANCE OF MATERIALS,

And an Appendix

ON THE

PRESERVATION OF TIMBER.

BY

DE VOLSON WOOD,

PROFESSOR OF CIVIL ENGINEERING IN THE UNIVERSITY OF MICHIGAN.

NEW YORK:
JOHN WILEY & SON,
15 ASTOR PLACE.
1871.

THE NEW YORK PRINTING COMPANY,
205, 207, 209, 211, *and* 213 *East Twelfth Street.*

PREFACE.

THIS book contains the substance of my lectures to the Senior Class in Civil Engineering, in the University, during the past few years, on the *Resistance of Materials*. The chief aim has been to present the theories as they exist at the present time. The subject is necessarily an experimental one, and any theory which has not the results of experiments for its foundation is valueless. I have therefore presented the results of a few experiments under each head, as they have been obtained in various parts of the world, that the student may judge for himself whether the theory is well founded or not. It is hoped that this part of the work will be valuable to the practical man.

The descriptive parts are given more fully here than they were in the lectures, because they can be consulted more profitably on the printed page than they could in the manuscript, and will be examined more by the general reader than the mathematical part. But, on the other hand, the mathematical part is much more condensed here than it was in the class-room. This was done so as to keep the work in as small a space as possible; and also because a student is supposed to have time for deliberate study, and can take time to overcome his difficulties and secure his results. It is intended, however, in the next edition, to publish an appendix, in order to explain the more difficult mathematical operations of the text.

I have taken special pains to make frequent references to other books and reports from which I have secured information. This will enable any one to verify more fully the positions which have been taken, and will be convenient for those who desire to secure a more thorough knowledge of any particular topic.

I do not deem it necessary to indicate those topics which are wholly original. To the reader who has never before given the subject any attention, all will be new; and the well-informed reader will readily detect what is original.

A large amount of labor and study has been given to this subject in nearly all civilized countries, and yet the theories in regard to resistance from transverse stress are not very satisfactory. In regard to the strength of rectangular beams, the "Common Theory," as I have called it, is sufficiently correct for ordinary practical purposes, especially if the *modulus of rupture*, as determined by direct experiment upon rectangular beams of the same material, be used. Barlow's "Theory of Flexure" appears to be more nearly correct in theory when applied to rectangular beams and beams of the **I** section, or other forms which are symmetrical in reference to the neutral axis. But when the sections are irregular none of the theories can be relied upon for securing correct results. Whatever theory may yet do for us, it is quite evident that no theory will ever be devised, of practical value, which will be applicable to the infinite variety of forms of beams which are or may be used in the mechanic arts. That I may not be misunderstood upon this point, I will be more specific. We know that our present theories do not always give correct results, and that the more irregular the form the greater the discrepancy between the actual and computed strengths of a beam. Now, if a theory is ever devised which will take into account all the conditions of strains in a beam, I think it will be too complicated to be of practical value to the mechanic. I do not desire by this remark to disparage theory. Theories are valuable. Without them we would make little or no progress. Fortunately for the engineer, it is not the mathematically exact result that he desires, but the *reliable* result. He does not so much desire to know that one pound more of load will break his structure, as he does that he may depend upon it to carry from four to six times the load which he intends to put upon it. The theories, as now developed, are safe guides to the mechanic and engineer; still we learn to depend more and more upon direct experiment. The theory also in regard to the deflection of beams under a transverse strain, has recently received a modification, due to a consideration of the effect of transverse shearing; but the modification is sustained both from mathematical and experimental considerations. May not more careful experiments yet teach us that it must be still further modified on account of the longitudinal shearing strain?

The author will be pleased to receive the results of experiments which have been made in this country, so that if this work is revised in the future, it may be made more profitable to the engineering profession.

ANN ARBOR, MICH., *Sept.*, 1871.

TABLE OF CONTENTS.

INTRODUCTION.

CHAPTER I.

TENSION.

CHAPTER II.

COMPRESSION.

CHAPTER III.

THEORIES OF FLEXURE AND RUPTURE FROM TRANSVERSE STRESS.

CHAPTER IV.

SHEARING STRESS.

CHAPTER V.

FLEXURE.

CHAPTER VI.

TRANSVERSE STRENGTH.

CHAPTER VII.

BEAMS OF UNIFORM RESISTANCE.

CHAPTER VIII.

TORSION.

CHAPTER IX.

EFFECTS OF LONG-CONTINUED STRAINS AND OF SHOCKS—CRYSTALLIZATION.

CHAPTER X.

LIMITS OF SAFE LOADING FOR MECHANICAL STRUCTURES.

APPENDIX I.

ON THE PRESERVATION OF WOOD.

APPENDIX II.

TABLE OF THE PROPERTIES OF MATERIALS.

A TREATISE

ON

THE RESISTANCE OF MATERIALS.

INTRODUCTION.

1. IN PROPORTIONING ANY MECHANICAL STRUCTURE, there are at least two general problems to be considered:—

1st. The nature and magnitude of the forces which are to be applied to the structure, such as moving loads, dead weights, force of the wind, etc.; and,

2d. The proper distribution and magnitude of the parts which are to compose the structure, so as to successfully resist the applied forces.

These problems are independent of each other. The former may be solved without any reference to the latter, as the structure may be considered as composed of rigid right lines. The latter depends principally upon the mechanical properties of the materials which compose the structure, such as their strength, stiffness, and elasticity, under various circumstances.

The mechanical properties of the principal materials—wood, stone, and iron *—have been determined with great care and expense by different experimenters, both in this and foreign countries, to which reference will hereafter be made.

---

* The properties of mortars have been thoroughly discussed by Gen. Q. A. Gilmore in his work on Limes, Mortars, and Cements. 1862.

1

2. DEFINITIONS OF TERMS.

STRESSES are the forces which are applied to bodies to bring into action their elastic and cohesive properties. These forces cause alterations of the forms of the bodies upon which they act.

STRAIN is a name given to the kind of alterations produced by the *stresses*. The distinction between stress and strain is not always observed; one being used for the other. One of the definitions given by lexicographers for *stress*, is *strain;* and inasmuch as the kind of distortion at once calls to mind the manner in which the force acts, it is not essential for our purpose that the distinction should always be made.

A TENSILE STRESS, or *Pull*, is a force which tends to elongate a piece, and produces a strain of extension, or *tensile strain*.

A COMPRESSIVE STRESS, or *Push*, tends to shorten the piece, and produces a *compressive strain*.

TRANSVERSE STRESS acts transversely to the piece, tending to bend it, and produces a *bending strain*. But as a compressive stress sometimes causes bending, we call the former a *transverse strain*, for it thus indicates the character of the stress which produces it. *Beams* are generally subjected to transverse strains.

TORSIVE STRESS causes a twisting of the body by acting tangentially, and produces a *torsive strain*.

LONGITUDINAL SHEARING STRESS, sometimes called a *detrusive strain*, acts longitudinally in a fibrous body, tending to draw one part of a solid substance over another part of it; as, for instance, in attempting to draw the piece A B, Fig. 1, which has a shoulder, through the mortise C, the part forming the shoulder will be forced longitudinally off from the body of the piece, so that the remaining part may be drawn through.

FIG. 1.

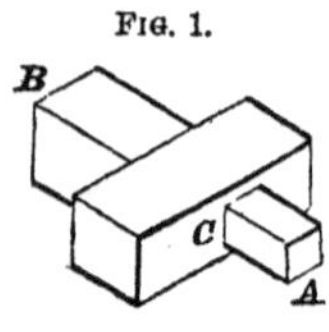

TRANSVERSE SHEARING STRESS is a force which acts transversely, tending to force one part of a solid body over the adjacent part. It acts like a pair of shears. It is the stress which would break a tenon from the body of a beam, by acting perpendicular to the side of the beam and close to the tenon. It is the stress which shears large bars of iron transversely, so often seen in machine-shops. The applied and resisting forces act in parallel planes, which are very near each other.

SPLITTING STRESS, as when the forces act normally like a wedge, tending to split the piece.

3. THE EFFECT OF THESE STRESSES IS TWOFOLD:— 1st. Within certain limits they only produce change of form; and, 2d, if they be sufficiently great they will produce rupture, or separation of the parts; and these two conditions give rise to two general problems under the resistance of materials, the former of which we shall call the problem of ELASTIC RESISTANCE; the latter, ULTIMATE RESISTANCE, or RESISTANCE TO RUPTURE.

4. GENERAL PRINCIPLES OF ELASTIC RESISTANCES.— To determine the laws of elasticity we must resort to experiment. Bars or rods of different materials have been subjected to different strains, and their effects carefully noted.

From such experiments, made on a great variety of materials, and with apparatus which enabled the experimenter to observe very minute changes, it has been found that, whatever be the physical structure of the materials, whether fibrous or granular, they possess certain general properties, among which are the following:—

1st. That all bodies are elastic, and within very small limits they may be considered perfectly elastic; *i. e.*, if the particles of a body be displaced any amount within these limits they will, when the displacing force is removed, return to the same position in the mass that they occupied before the displacement. This limit is called *the limit of perfect elasticity.**

* Mr. Hodgkinson made some experiments to prove that all bodies are non-elastic. (See *Civil Eng. and Arch. Jour.* vol. vi., p. 354.) He found that the limits of perfect elasticity were exceedingly small, and inferred that if our

2d. The amount of displacement within the elastic limit is directly proportional to the force which produces it. It follows from this, that in any prismatic bar the force which produces compression or extension, divided by the amount of extension or compression, will be a constant quantity.

3d. If the displacement be carried a little beyond this limit the particles will not return to their former position when the displacing force is removed, but a part or all of the displacement will be permanent. This Mr. Hodgkinson called a *set*, a term which is now used by all writers upon this subject.

4th. The amount of displacement is not exactly, but nearly, proportional to the applied force considerably beyond the elastic limit.

5th. Great strains, producing great sets, impair the elasticity.

5. COEFFICIENT (OR MODULUS*) OF ELASTICITY.

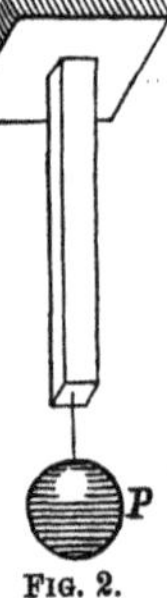

FIG. 2.

If a prismatic bar, whose section and length are unity, be compressed or elongated any amount within the elastic limit, the quotient obtained by dividing the force which produces the displacement by the amount of compression or extension is called the COEFFICIENT OF ELASTICITY. This we call E. Let K=section of a prismatic bar (See Fig. 2), l=its length,

---

powers of observation were perfect in kind and infinite in degree, we should find that no body was perfectly elastic even for the smallest amount of displacement. And although more recent experiments have *indicated* the same result in cast-iron, yet the most delicate experiments have failed to thoroughly establish it. I have, therefore, accepted the principle of perfect elasticity, which, for the purposes of this work, is practically, if not theoretically, correct. It does not appear from Mr. Hodgkinson's report how soon the effect was observed after the strain was removed. If he had allowed considerable time the *set* might have disappeared, as it is evident that it takes time for the displaced particles to return to their original position.

* The terms *coefficient* and *modulus* are used indiscriminately for the constants which enter equations in the discussion of physical problems, and are sometimes called *physical constants.* The *modulus* of elasticity, as used by most writers on Analytical Mechanics, is the ratio of the force of restitution to

and λ=the elongation or compression caused by a force, P, which is applied longitudinally. Then

$\frac{P}{K}$=force on a unit of section, and

$\frac{\lambda}{l}$=the elongation or compression for a unit of length.

Hence, from the definition given above, we have

$$E = \frac{P}{K} \div \frac{\lambda}{l} = \frac{Pl}{K\lambda} \quad - \ - \ - \ - \ - \ - \ - \quad (1)$$

From this equation E may be easily found. It will hereafter be shown that the coefficient is not exactly, but is nearly the same for compression as for tension.

For values of E, see Appendix, Table 1.

6. PROOFS OF THE LAWS GIVEN IN ARTICLE FOUR.— Article 5 has preceded these proofs, so as to show how the results of experiments may be reduced by equation (1). The 1st and 2d laws seem first to have been proved by S. Gravesend, since which they have been confirmed by numerous experimenters. One of the most extensive and reliable series of experiments upon various substances for engineering purposes is given in "The Report of Her Majesty's Commissioners, made under the direction of Mr. Eaton Hodgkinson." The results of his experiments are published in the Reports of the British Association, and in the 5th volume of the Proceedings of the Manchester Literary and Philosophical Society, from which extracts have been made and to which we shall have occasion to refer. The experiments were made not only to prove these laws but several others, principally relating to transverse strength.

Barlow made many experiments, the results of which are given in his valuable work on the "Strength of Materials." The series of experiments on iron which had been commenced and so ably

---

that of compression. It relates to the impact of bodies, and, as thus defined, depends upon the set. But the *coefficient* of elasticity depends neither upon impact nor set. Another term should therefore be used, or else a distinction should be made between the terms *coefficient* and *modulus*, so that the former shall apply to small displacements, and the latter to the relative force of restitution. For this reason I have used the former in this work, and avoided the latter when applied to elasticity.

conducted by Mr. Hodgkinson were continued by Mr. Fairbairn. The latter confined his experiments mostly to transverse strength, the results of which are given in his valuable work on "Cast and Wrought Iron." A valuable set of experiments has been made in France at "le Conservatoire des Arts et Métiers." *

In this country several very valuable sets of experiments have been made, among the most important of which are the experiments of the Sub-Committee of the Franklin Institute, the results of which are published in the 19th and 20th volumes of the Journal of that Society, commencing on the 73d page of the former volume. The experiments were made upon boiler iron, but they developed many properties common to all wrought iron. They were conducted with great care and scientific skill. The report gives a description of the testing machine; the manner of determining its friction and elasticity; the modifications for use in high temperature; the manner of determining the latent and specific heats of iron; and the strength of different metals under a variety of circumstances.

Another very valuable set of experiments was made by Captain T. J. Rodman and Major W. Wade, upon "Metals for Cannon, under the direction of the United States Ordnance Department," and published by order of the Secretary of War.

Numerous other experiments of a limited character have been made, too many of which have been lost to science because they were not reported to scientific journals, and many others were of too rude a character to be very valuable.

The results of these experiments will form the basis of our theories and analysis.

* See "Morin's Résistance des Matériaux," p. 126.

CHAPTER I.

TENSION.

7. TAKING THE PHENOMENA IN THEIR NATURAL ORDER, the first thing which claims attention is the elastic resistance due to tension, or, as it is sometimes called, a pull, or elongating force.

EXPERIMENTS ON WROUGHT IRON.

Experiments for determining the total elongation and permanent elongation produced by different weights acting by extension on a tie of wrought iron of the best quality, by Eaton Hodgkinson.

Weight in kilogrammes per square centimetre. P.	Elongation per metre of length.		Coefficient of elasticity per square metre. E.
	Total. λ.	Permanent.	
Kil.	M.	Mill.	Kil.
187.429	0.000082117	....	22 824 500 000
374.930	0.000185261	....	20 216 200 000
562.406	0.000283704	0.00254	19 824 100 000
749.456	0.000379467	0.0033894	19 704 000 000
937.430	0.000475113	0.0042398	19 729 909 000
1124.813	0.000570702	0.00508	19 706 000 000
1312.283	0.000665647	0.0067705	19 714 600 000
1499.720	0.000760311	0.0100879	19 320 300 000
1687.219	0.000873265	0.0330283	19 320 700 000
1874.645	0.001012911	0.0829955	18 398 100 000
2063.580	0.001283361	0.2616950	16 079 200 000
2249.627	0.002227205	.....	.. ..
2403.653	0.004287185	3.0709900	5 606 590 000
2624.564	0.009156490	8.4690700	2 866 380 000
....	0.009950970	8.5748700	.. ..
2812.033	0.010492805	9.1023600	2 681 520 000
Repeated after 1 hour.	0.011750313	....	.. ..
" " 2 "	0.011858889	....	.. ..
" " 3 "	0.011933837	....	.. ..
" " 4 "	0.011942168	....	.. ..
" " 5 "	0.011958835	....	.. ..
" " 6 "	0.011967149	....	.. ..
" " 7 "	0.012027114	....	.. ..
" " 8 "	0.012027014	....	.. ..

EXPERIMENTS ON WROUGHT IRON.—*Continued.*

Weight in kilogrammes per square centimetre. P.	Elongation per metre of length.		Coefficient of elasticity per square metre. E.
	Total. λ.	Permanent.	
Kil.	M.	Mill.	Kil.
Repeated after 9 hours.	0.012027114	....	... ...
" " 10 "	0.012027114	....	... ...
2999.500	0.017888263	16.5145	1 676 820 000
2999.500	0.019478898	....	... ...
....	0.01984831	18.4212	... ...
....	0.02022006	18.8886	... ...
3186.975	0.02148590	19.7954	1 483 290 000
....	0.02169401	....	... ...
....	0.02170242	....	... ...
....	0.02170242	22.0119	... ...
3374.440	0.02477441	22.7087	1 362 020 000
....	0.02514184	....	... ...
....	0.02522512	....	... ...
3561.900	0.03493542	32.8201	1 019 580 000
....	0.03519357	....	... ...
....	0.03520190	....	... ...
3745.361	....	....	... ...

This table is given in French units because it was more convenient.*

8. THE RESULTS OF THESE EXPERIMENTS may be represented graphically by taking, as has been done in Fig. 3, the total elongations or the permanent elongations for abscissas and the weights for ordinates.

---

* To reduce the French measures to English we have the following relations:—

LINEAR MEASURE.

3.2808992 feet = 1 metre.
0.0328089 feet = 1 centimetre.
0.0032808 feet = 1 millimetre.
0.0393696 in. = 1 millimetre.

WEIGHT.

2.20462 lbs. avoird. = 1 kilogramme.
1422.28 lbs. pr. sq. in. = 1 kilog. to the sq. millimetre.
0.00142228 lbs. sq. in. = 1 kilog. pr. sq. metre.

Hence to reduce the above quantities to English units, multiply the numbers in the first column by 14.2228 to reduce them to pounds avoirdupois per square inch; those in the second column by 3.28089 + to reduce them to feet; the third by 0.03936 + to reduce them to inches; and the fourth by 0.00142228 to reduce them to pounds per square inch.

When the construction is made on a large scale it makes the results of the experiments very evident.

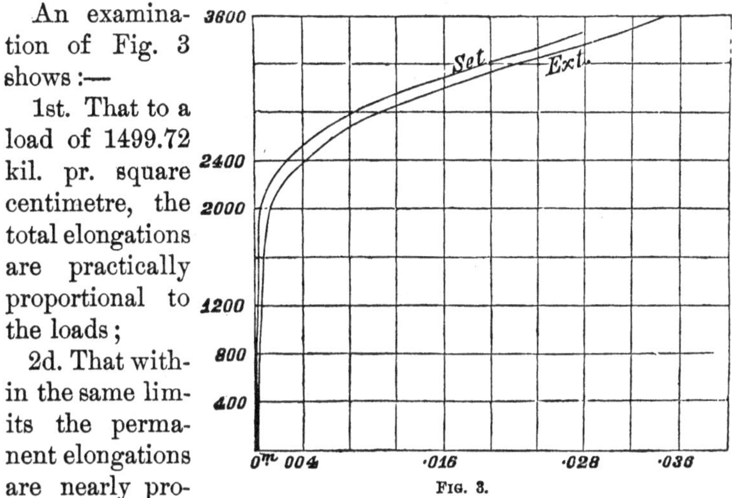

Fig. 3.

An examination of Fig. 3 shows:—

1st. That to a load of 1499.72 kil. pr. square centimetre, the total elongations are practically proportional to the loads;

2d. That within the same limits the permanent elongations are nearly proportional to the loads, and that they are exceedingly small;

3d. That beyond the load of 14.997 kil. to 22.00 kil. per square millimetre, the total and permanent elongations increase very rapidly and more than proportional to the loads;

4th. That near and beyond 22.49 kil. per square millimetre, the total elongations become sensibly proportional to the loads, but in a much greater ratio than that which corresponds to small loads. For the loads near rupture the elongations are a little inferior to that indicated by the new proportion.

5th. Beyond 14.99 kil. per square millimetre, the permanent elongations increase much more rapidly than the total elongations. We also observe that the permanent elongations increase with the duration of the load, although very slowly. The latter property will be more particularly noticed hereafter.

6th. Finally, the values $\frac{P}{\lambda}$ of the loads per square metre to the elongation per metre, and which is called the *coefficient of elasticity*, is sensibly constant when the elongations are nearly proportional to the loads; and that the mean value is

$$E = 19{,}816{,}440{,}000 \text{ kil. per square metre;}$$
$$= 28{,}283{,}000 \text{ lbs. per square inch.}$$

The first value of E, in the table, is much larger, and may

have resulted from an erroneous measurement of the exceedingly small total elongation. From the experiments made on another bar, Hodgkinson found

$$E = 19,359,458,500 \text{ kil. per sq. metre;}$$
$$= 27,700,000 \text{ pounds pr. square inch;}$$

which is but little less than the preceding.

Mr. Hodgkinson infers from these experiments that the smallest strains cause a permanent elongation. But Morin forcibly remarks * that none of these experimenters appear to have verified whether time, after the strains are removed, will not cause the permanent elongations to disappear. Also that the deflections of the machine cannot be wholly eliminated, and hence appear to increase the true result. In practice such small permanent elongations may be omitted.

The preceding example has, for a long time, been given to show the law of relation between the applied force and the total and permanent elongations; but we should not expect to find exactly the same results for all kinds of iron. Even wrought iron has such a variety of qualities, depending upon the ore of which it is made, and the process of manufacture, that it cannot be expected that the above results will always be applicable to it. Only a wide range of experiments will determine how far they may generally be relied upon.

It is found, however, that the GENERAL RESULTS of extension, of set, of increased elongation with the duration of the stress within certain limits, and of the increase of set with the increase of load, are true of all kinds of iron.

EXPERIMENTS UPON CAST IRON.

9. THE FOLLOWING EXPERIMENTS UPON CAST IRON show that the numerical relation between the applied force and the extension is somewhat different from the preceding. The experiments were made under the supervision of Captain T. J. Rodman:— †

"The specimens had collars left on them at a distance of thirty-five inches

---

* Morin's Résistance des Matériaux, p. 10.

† Experiments on Metals for Cannon, by Capt. T. J. Rodman, p. 157.

For a full description of the testing apparatus, with diagrams, see Major Wade's Report on the Strength of Materials for Cannon, pp. 305–315. The machine consists principally of a very substantial frame and levers resting on knife edges.

apart, the space between the collars being accurately turned throughout to a uniform diameter.

"The space between the collars was surrounded by a cast-iron sheath, eight-tenths of an inch less in length than the distance between the collars; it was put on in halves and held in position by bands, and was of sufficient interior diameter to move freely on the specimen.

"When in position, the lower end of the sheath rested on the lower collar of the specimen, the space between its upper end and the upper collar being supplied with and accurately measured by a graduated scale tapered 0.01 of an inch to one inch.

"The upper end of the sheath was mounted with a vernier, and the scale was graduated to the tenth of an inch.

"This afforded means of measuring the changes of distance between the collars to the ten-thousandth part of an inch, and these readings divided by the distance between the collars gave the extension per inch in length as recorded in the following table :—

TABLE

Showing the extension and permanent set per inch in length caused by the under-mentioned weights, per square inch of section, acting upon a solid cylinder 35 *inches long and* 1.366 *inches diameter.* (*Cast at the West Point Foundry in* 1857.)

Weight per square inch of section.	Extension per inch of length.	Permanent set per inch in length.	Coefficient of elasticity.
P.	λ.		E.
lbs.	in	in.	
1,000	0.0000611	0.	16,366,612
2,000	0.0000794	0.	25,189,168
3,000	0.0001089	0.	27,548,209
4,000	0.0001771	0.	22,586,674
5,000	0.0002129	0.	23,489,901
6,000	0.0002700	0.0000014	22,222,222
7,000	0.0003328	0.0000029	21,033,653
8,000	0.0003986	0.0000043	20,070,245
9,000	0.0004557	0.0000071	19,749,835
10,000	0.0005100	0.0000109	19,607,843
11,000	0.0005500	0.0000157	20,000,000
12,000	0.0006414	0.0000257	18,693,486
13,000	0.0007100	0.0000300	18,309,859
14,000	0.0007700	0.0000357	18,181,181
15,000	0.0008557	0.0000477	17,529,507
16,000	0.0009243	0.0000529	17,310,397
17,000	0.0010014	0.0000643	16,977,231
18,000	0.0010900	0.0001014	16,537,614
19,000	0.0012271	0.0001471	15,483,660
20,000	0.0013586	0.0002014	14,721,109
21,000	0.0015386	0.0002900	13,648,771
22,000	0.0017043	0.0003986	12,908,523
23,000	0.0019529	0.0005529	11,265,246
24,000	0.0022786	0.0007529	10,532,344
25,000	0.0026037	0.0010843	9,601,720
26,000	0.0032186		8,078,046

10. FIGURE 4 IS A GRAPHICAL REPRESENTATION OF THE ABOVE TABLE, constructed in the same way as Figure 3.

Experiments were made upon many other pieces, from which I have selected four, and called them A, B, C, and D, a graphical representation of which is shown in Figure 5. The right hand lines represent extensions, the left hand sets.

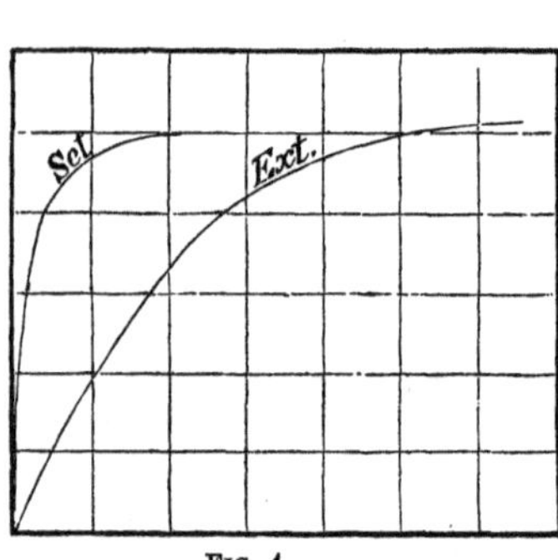

FIG. 4.

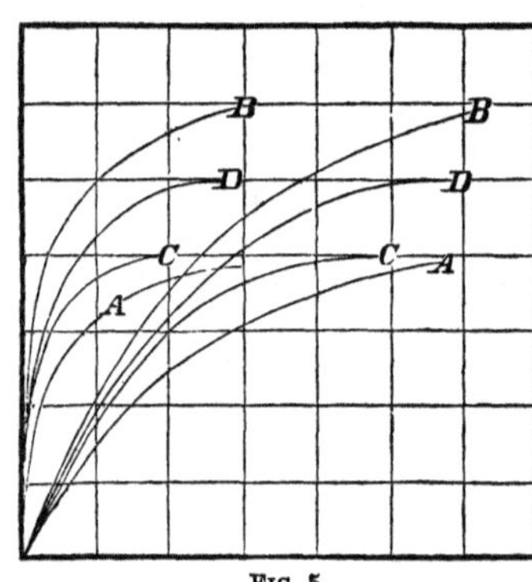

FIG. 5.

A is from an inner specimen of a Fort Pitt gun, No. 335, and the others from different cylinders which were cast for the purpose of testing the iron.

From these we observe :—

1st. That for small elongations the ratio of the stresses to the elongations is nearly constant.

2d. There does not appear to be a sudden change of the rate of increase, as in Mr. Hodgkinson's example, but the ratio gradually increases as the strains increase.

3d. The sets at first are invisible, but they increase rapidly as the strains approach the breaking limit.

It appears *paradoxical* that the first and second experiments in the preceding table should give a less coefficient than the third, but the same result was observed in several cases.

11. THE FOLLOWING TABLES ARE THE RESULTS OF SOME EXPERIMENTS MADE BY MR. HODGKINSON :—

Direct longitudinal extension of round rods of cast iron, fifty feet long.

NAME OF IRON.	No. of experiments.	Mean area of section.	Weights per square inch laid on, with their corresponding extensions and sets.			Mean breaking weight per square inch of section.	Mean ultimate extension.
			Weights.	Extension.	Sets.		
			lbs.	in.	in.	lbs.	in.
Low Moor, No. 2......	2	1.058	2,117	0.0950	0.00345	16,408	1.085
			6,352	0.3115	0.0250		
			10,586	0.5740	0.4425		
			14,821	0.9147	0.12775		
Blaneavon Iron, No. 2..	2	1.0685	2,096	0.0942	0.00268	14,675	0.9325
			6,289	0.3065	0.01675		
			10,482	0.5770	0.0575		
			13,627	0.8370	0.11475		
Gartsherrie Iron, No. 2.	2	1.062	2,109	0.0922	0.001 +	16,951	1.167
			6,328	0.3117	0.01450		
			10,547	0.5862	0.0475		
			14,766	0.9452	0.11352		

In these experiments the ratio of the extensions is somewhat greater than that of the weights. The value of E, as computed for the first weights which are given, and the corresponding extensions, is a little more than 13,000,000 pounds per square inch.

Extension of cast-iron rods, ten feet long and one inch square.				Error in parts of the weight when it is computed from formula $P=1161r\text{-}\lambda_e-201905\lambda^2$.
Weights P.	Extensions. λ_e.	Sets.	$\frac{P}{\lambda_e}$	
lbs.	in.	in.		
1053.77	.0090	.00022	117086	$-\frac{1}{42}$
1580.65	.0137	.000545	115131	$-\frac{1}{37}$
2167.54	.0186	.00107	113309	$-\frac{1}{119}$
3161.31	.0287	.00175	110150	$+\frac{1}{640}$
4215.08	.0391	.00265	107803	$+\frac{1}{257}$
5262.85	.0500	.00372	105377	$+\frac{1}{163}$
6322.62	.0613	.00517	103142	$+\frac{1}{177}$
7376.39	.0734	.00664	100496	$+\frac{1}{125}$
8430.16	.0859	.00844	98139	$+\frac{1}{155}$
9483.94	.0995	.01062	95316	$+\frac{1}{134}$
10537.71	.1136	.01306	92762	$+\frac{1}{121}$
11591.48	.1283	.01609	90347	$-\frac{1}{674}$
12645.25	.1448	.02097	87329	$-\frac{1}{195}$
13699.83	.1668	.	82133	$+\frac{1}{268}$
14793.10	.1859	.02410	79576	$-\frac{1}{80}$

Let P = the elongating force and

λ_e = the total elongation in inches due to P.

Then Hodgkinson found, from an examination of the table, that the empirical formula

$$P = 116117\lambda_e - 201905\lambda_e^2.$$

represented the results more nearly than equation (1). This formula reduced to an equivalent one for l in inches (observing that the bar was 10 feet long), becomes

$$P = 13{,}934{,}000\ \frac{\lambda_e}{l} = 2{,}907{,}432{,}000 \frac{\lambda_e}{l^2}$$

Although this equation gives the elongations for a greater range of strains than equation (1) for this particular case, yet the law represented by it is more complicated, and hence would make the discussions under it more difficult, without yielding any corresponding advantage. It is the equation of a parabola in which P is the abscissa and λ_e the ordinate.

We also see that when the elongations are very small, the quantity $\frac{\lambda_e^2}{l^2}$ will be very small, and the second term may be omitted in comparison with the first, in which case it will be reduced to equation (1). The coefficient in the first term is the coefficient of elasticity, hence it is nearly 14,000,000 lbs. for extension.

MALLEABLE IRON.

12. According to Barlow's experiments malleable iron may be elongated $\frac{1}{1000}$ of its length without endangering its elasticity.* To ascertain this, the strains were removed from time to time, and it was found that the index returned to zero for all strains less than 9 or 10 tons. The mean extension per ton (of 2,240 lbs.) per square inch, for four experiments, was 0.00009565 of its original length. Hence the mean value of the coefficient of elasticity is

$$E = \frac{P}{\lambda} = \frac{2240}{0.00009565} = 23{,}418{,}000 \text{ lbs.}$$

* Journal Frank. Inst., vol. xvi., 2d Series, p. 126.

ELASTICITY OF WOOD.

13. EXPERIMENTS BY MESSRS. CHEVANDIER AND WERTHEIM.—The following are some of the results of the recent experiments of Messrs. Chevandier and Wertheim on the resistance of wood. These experimenters have drawn the following principal conclusions:—

1. The density of wood appears to vary very little with age.
2. The coefficient of elasticity diminishes, on the contrary, beyond a certain age; it depends, likewise, upon the dryness and the exposure of the soil to the sun in which the trees have grown; thus the trees grown in the northern exposures, north-eastern, north-western, and in dry soils, have always so much the higher coefficient as these two conditions are united, whereas the trees grown in muddy soils present lower coefficients.
3. Age and exposure influence cohesion.
4. The coefficient of elasticity is affected by the soil in which the tree grows.
5. Trees cut in full sap, and those cut before the sap, have not presented any sensible differences in relation to elasticity.
6. The thickness of the woody layers of the wood appeared to have some influence on the value of the coefficient of elasticity only for fir, which was greater as the layers were thinner.
7. In wood there is not, properly speaking, any limit of elasticity, as every elongation produces a set.

It follows from this circumstance that there is no limit of elasticity for the woods experimented upon by Messrs. Chevandier and Wertheim, but in order to make the results of their experiments agree with those of their predecessors, the authors have given for the value of the limit of elasticity the load under which it produces only a very small permanent elongation; the limit which they indicate in the following table for loads under which the elasticity of wood commences to change, corresponds to a permanent elongation of 0.00005, its original length.

TABLE CONTAINING THE MEAN RESULTS OF THE EXPERIMENTS OF MESSRS. CHEVANDIER AND WERTHEIM.

Species.	Density.	Coefficient of elasticity E referred to the square millimetre.	Limit of elasticity, or load per square millimetre, corresponding to that limit.	Cohesion, or load per square millimetre capable of producing rupture.	Val. of E in pounds per square inch.
		Kilogr.	Kilogr.	Kilogr.	
Locust..................	0.717	1261.9	3.188	7.93	The coefficient of elasticity varies from 1,000,000 lbs. to nearly 1,800,000 lbs. per square inch.
Fir.....................	0.493	1113.2	2.153	4.18	
Yoke Elm...............	0.756	1085.3	1.282	2.99	
Birch..................	0.812	997.2	1.617	4.30	
Beech	0.823	980.4	2.317	3.57	
Oak from pedunculate acorn	0.808	977.8	"	6.49	
" " sessile acorn....	0.872	921.8	2.349	5.66	
White Pine..............	0.559	564.1	1.633	2.48	
Elm	0.723	1165.3	1.842	6.99	
Sycamore................	0.692	1163.8	1.139	6.16	
Ash.....................	0.697	1121.4	1.246	6.78	
Alder...................	0.601	1108.1	1.121	4.54	
Aspen...................	0.602	1075.9	1.035	7.20	
Maple...................	0.674	1021.4	1.068	3.58	
Poplar..................	0.477	517.2	1.007	1.97	

14. ELASTICITY OF WOOD, TANGENTIALLY AND RATIALLY.—The same observers have also determined the coefficient of elasticity and the cohesion of wood in the direction of the radius and in the direction of the tangent to the woody layers.

An examination of the following Table shows that the resistance in the direction of the radius is always greater than the resistance in the direction of the tangent to the woody layers; the relation between the coefficients of elasticity in the two cases varying nearly from 3 to 1.15.

MEAN RESULTS OF THE EXPERIMENTS OF MESSRS. CHEVANDIER AND WERTHEIM.

SPECIES.	IN THE DIRECTION OF RADIUS.		IN THE DIRECTION OF THE TANGENT TO THE LAYERS.	
	Coefficient of Elasticity, E, per square millimetre.	Cohesion, or load, per square millimetre, capable of producing rupture.	Coefficient of Elasticity, E, per square millimetre.	Cohesion, or load, per square millimetre, capable of producing rupture.
	Kilogr.	Kilogr.	Kilogr.	Kilogr.
Yoke Elm............	208.4	1.007	103.4	0.608
Sycamore............	134.9	0.522	80.5	0.610
Maple...............	157.1	0.716	72.7	0.371
Oak.................	188.3	0.582	129.8	0.406
Birch...............	81.1	0.823	155.2	1.063
Beech...............	269.7	0.885	159.3	0.752
Ash.................	111.3	0.218	102.0	0.408
Elm.................	121.6	0.345	63.4	0.366
Fir.................	94.5	0.220	34.1	0.297
Pine................	97.7	0.256	28.6	0.196
Locust..............	170.3	"	152.2	1.231

The highest coefficient of elasticity in this table is for beech, and this is less than 400,000 pounds per square inch.

15. REMARK.—*The value of* E, which is used in practice, is not the coefficient of *perfect* elasticity, but it is that value which is nearly constant for small strains. In determining it, no account is made of the set. If the total elongations were proportional to the stresses which produce them, we would use the value of E found by them, even if the permanent equalled the total elongations. But in practice the permanent elongations will be small compared with the total for small stresses.

APPLICATIONS.

16. TO FIND THE ELONGATION OF A PRISMATIC BAR SUBJECTED TO A LONGITUDINAL STRAIN WHICH IS WITHIN THE ELASTIC LIMITS.

From (1) we have

$$\lambda = \frac{Pl}{EK} \qquad (2)$$

which is the required formula.

Also from (1) we have

$$P = \frac{\lambda}{l} EK \quad - \quad - \quad - \quad - \quad - \quad - \quad - \quad - \quad (3)$$

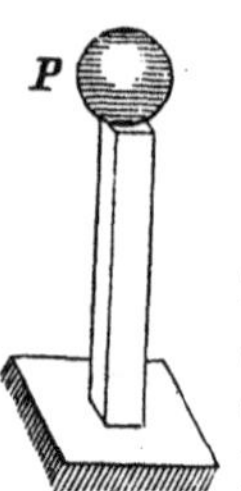

FIG. 6.

Equations (1), (2), and (3) are equally applicable to compressive strains, as will hereafter be shown. If in (3) we make $K=1$ and $\lambda=l$ we shall have $P = E$; hence, the *coefficient of elasticity* may be defined to be *a force which will elongate a bar whose section is unity, to double its original length, provided the elasticity of the material does not change.* But there is no material, not even a perfectly elastic body,—as air and other gases,—whose *coefficient* of elasticity will not change for a perceptible change of volume. The material may not lose its elasticity, but equation (1) only measures it for small displacements. To illustrate further, let it be observed that, according to Mariotte's law, the volumes of a gas are inversely proportional to the compressive (or extensive) forces; double the force producing a compression of half the volume; four times the force, one-fourth the volume, and so on,—the compressions being a *fractional* part of the original volume; but in equation (2), λ is a linear quantity, so that if one pound produces an extension (or compression) of one inch, two pounds would produce an extension of two inches, and so on.

Examples.—1. If the coefficient of elasticity of iron be 25,000,000 lbs., what must be the section of an iron bar 60 feet long, so that a weight of 5,000 lbs. shall elongate it $\frac{1}{2}$ inch?

From (1) we obtain $K = \frac{Pl}{E\lambda}$ which by substitution becomes

$$K = \frac{5,000.12 \times 60}{25,000,000 \times \frac{1}{2}} = 0.288 \text{ square inches.}$$

2. How great a weight will a brass wire sustain, whose diameter is 1 inch; coefficient of elasticity is 14,000,000 lbs., without elongating it more than $\frac{1}{800}$ of its length? Ans. 13,744.5 lbs.

17. REQUIRED THE ELONGATION (OR COMPRESSION) OF A PRISMATIC BAR WHEN ITS WEIGHT IS CONSIDERED.

Let $l =$ the whole length of the bar before elongation or compression,

$x =$ variable distance $= \mathrm{A}b$,

$dx = bc =$ an element of length,

$w =$ weight of a unit of length of the bar,

$\mathrm{W} =$ weight of the bar, and

$\mathrm{P}_1 =$ the weight sustained by the bar.

Then $(l - x)\,w + \mathrm{P}_1 = \mathrm{P} =$ the strain on any section, as bc.

Fig. 7.

Hence, from equation (2), we have

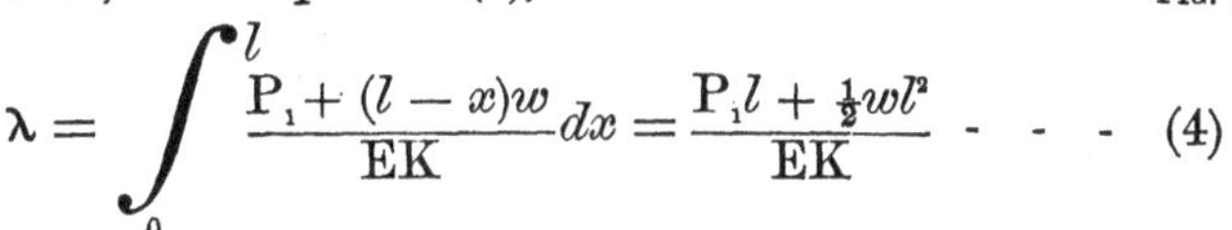

$$\lambda = \int_0^l \frac{\mathrm{P}_1 + (l - x)w}{\mathrm{EK}}\,dx = \frac{\mathrm{P}_1 l + \frac{1}{2}wl^2}{\mathrm{EK}} \quad - \ - \ - \quad (4)$$

$\therefore$ the total length will become,

$$l \pm \lambda = \left[1 \pm \frac{\mathrm{P}_1 + \frac{1}{2}wl}{\mathrm{EK}}\right] l \quad - \ - \ - \ - \ - \ - \ - \ - \quad (5)$$

If $\mathrm{P}_1 = 0$, $\lambda = \dfrac{wl^2}{2\mathrm{EK}} = \dfrac{\mathrm{W}l}{2\mathrm{EK}}$, or the total elongation is one-half of what it would be if a weight equal to the whole weight of the bar were concentrated at the lower end.

Required the Elongation (or Compression) of a Cone in a Vertical Position, caused by its own Weight when it is suspended at its Base (or rests on its Base).

Take the origin at the apex before extension, Fig. 8, and

let $\mathrm{K} =$ any section,

$\mathrm{K}_0 =$ the upper section,

$l =$ the length or altitude of the cone,

$x =$ the length or altitude of any portion of the cone, and

$\delta =$ the weight of a unit of volume.

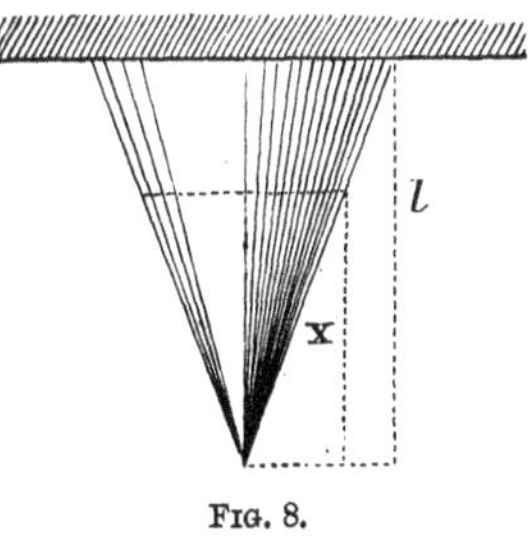

Fig. 8.

Then, because the bases of similar cones are as the squares of their altitudes, $\mathrm{K} = \mathrm{K}_0 \dfrac{x^2}{l^2}$

The volume of the cone whose altitude is x

$$=\int_0^x \mathrm{K}dx = \int_0^x \mathrm{K}_0 \frac{x^2}{l^2}dx = \tfrac{1}{3}\mathrm{K}_0 \frac{x^3}{l^2},$$

and the weight of the same part

$$= \tfrac{1}{3}\,\delta\,\mathrm{K}_0 \frac{x^3}{l^2}$$

$$\therefore \text{(from equation (2))}\ \lambda = \int_0^l \frac{\frac{1}{3}\frac{\delta \mathrm{K}_0}{l^2}x^3}{\mathrm{EK}_0\frac{x^2}{l^2}}dx = \frac{1}{6}\frac{\delta l^2}{\mathrm{E}}$$

from which it appears that the total elongation is independent of the transverse section, and varies as the square of the length.

18. The work of elongation.—If P be the force which does the work, and x the space over which it works, then the general expression for the work is

$$\mathrm{U} = \int_0^x \mathrm{P}dx \quad \text{- - - - - - - - - - - - - -} \quad (6)$$

To apply this to the prism, substitute P from Eq. (3) in (6), and make $dx = d\lambda$, and we have

$$\mathrm{U} = \int_0^\lambda \frac{\mathrm{EK}\lambda}{l}d\lambda = \frac{\mathrm{EK}\lambda^2}{2l} = \tfrac{1}{2}\mathrm{P}\lambda \quad \text{- - - - - -} \quad (7)$$

which is the same result that we would have found by supposing that P was put up by increments, increasing the load gradually from zero to P.

Example.—If the coefficient of elasticity of wrought iron be 28,000,000 lbs., and is expanded 0.00000698 of its length for one degree F., how much work is done upon a prismatic bar whose section is one inch, and length 30 feet, by a change of 20 degrees of temperature?

Walls of buildings which were sprung outward have been drawn into an erect position by heating and cooling bars of iron. Several rods were passed through the buildings, and extending from wall to wall, were drawn tight by means of the nuts. Then a part of them were heated, thus elongating them, and the nuts tightened; after which they were allowed to cool, and the contraction which resulted drew the walls together. Then the other rods were treated in a similar manner, and so on alternately.

19. VERTICAL OSCILLATIONS.—If a bar Aa, Fig. 9, with a weight, P, suspended from its lower end, be pressed down by the hand, or by an additional weight from a to b, and the additional force be suddenly removed, the end of the bar on returning will not stop at a, but will move to some point above, as c, a distance $ac = ab$. From a principle in Mechanics, viz., that the *living force* equals twice the work, we are enabled to determine all the circumstances of the oscillation when the weight of the bar is neglected. The weight P elongates the bar so that its lower extremity is at a, at which point we will take the origin of co-ordinates.

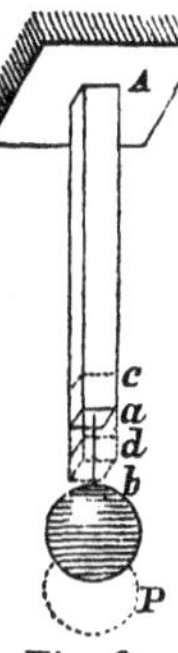

Fig. 9.

Let $\lambda = ab =$ the elongation caused by the additional force,
$x = ad =$ any variable distance from the origin,
$v =$ the velocity at any point, as d, and
$\text{M} =$ the mass of the weight P.

If the weight of the rod be very small compared with P, the *vis viva* is

$$\text{M}v^2 = \frac{\text{P}}{g}v^2 \text{ very nearly.}$$

The work for an elongation equal to λ, is by Eq. (7), $\frac{\text{EK}}{2l}\lambda^2$

" " " x " " $\frac{\text{EK}}{2l}x^2$

$$\therefore \tfrac{1}{2}\frac{\text{P}}{g}v^2 = \frac{\text{EK}}{2l}(\lambda^2-x^2), \text{ or } \frac{dx^2}{dt^2} = g\frac{\text{EK}}{\text{P}l}(\lambda^2-x^2) = v^2$$

$$\therefore t = \sqrt{\frac{\text{P}l}{g\text{EK}}}\int_0^\lambda \frac{dx}{\sqrt{\lambda^2-x^2}} = \sqrt{\frac{\text{P}l}{g\text{EK}}}\sin^{-1}\frac{x}{\lambda}\Big]_0^\lambda = \frac{\pi}{2}\sqrt{\frac{\text{P}l}{g\text{EK}}}$$

for half an oscillation; and the time for a whole oscillation is

$$\text{T} = \pi\sqrt{\frac{\text{P}l}{g\text{EK}}} = \pi\sqrt{\frac{\lambda}{g}}, \qquad (8)$$

hence the oscillations will be isochronous.

It is evident that by applying and removing the force at regular intervals, the amplitude of the oscillations may be increased and possibly produce rupture. In this way the Broughton suspension bridge was broken.*

As a second example take the case in which P is applied suddenly to the end of the rod. It is evident that the total elongation will be greater than λ,—the permanent elongation. For the fundamental equation we may use another

* Mr. E. Hodgkinson, in the 4th volume of the *Manchester Philosophical Transactions*, gives the circumstances of the failure, from this cause, of the suspension bridge at Broughton, near Manchester, England. And M. Navier, in his theory of suspension bridges (*Ponts Suspendus*, Paris, 1823), states that the duration of the oscillation of chain bridges may be nearly six seconds.

principle in Mechanics, which might have been used in the preceding problem, viz., that the mass multiplied by the acceleration equals the moving force. The resisting force for an elongation x is $\frac{\text{EK}x}{l}$ (See Eq. (8)), and the moving force is P, whose mass $= \frac{\text{P}}{g}$; hence

$$\text{M}\frac{d^2x}{d^2t} = \text{P} - \frac{\text{EK}}{l}x;$$

$$\therefore \frac{dx^2}{dt^2} = \frac{g}{\text{P}}\left(2\text{P}x - \frac{\text{EK}}{l}x^2\right) = v^2$$

$$\therefore t = \sqrt{\frac{\text{P}}{g}}\int_0^x \frac{dx}{\sqrt{2\text{P}x - \frac{\text{EK}x^2}{l}}} = \sqrt{\frac{\text{P}}{g}\cdot\frac{l}{\text{EK}}}\ \text{versin}^{-1}\frac{\text{EK}x}{\text{P}l}$$

If $x = \lambda$, $v = \sqrt{g\lambda}$,
$x = 2\lambda$, $v = 0$,
$x = 0$, $v = 0$.

Hence, the amplitude is twice the permanent elongation. If $x = 2\lambda$ we have $t = \pi\sqrt{\frac{\text{P}l}{g\text{EK}}} = \pi\sqrt{\frac{\lambda}{g}}$. Investigations of this kind give rise to a division of the subject called *Resiliance of Prisms*.

The investigations are interesting, but the results are of little use beyond those which have already been indicated. From the last problem we see that a weight suddenly applied produces twice the strain that it would if applied gradually.

As additional exercises for the student, I suggest the following: Suppose the weight be applied with an initial velocity. Suppose a weight P is attached to one end, and the weight P′ is placed suddenly upon it; or it falls upon it. To find the velocity at any point in terms of t, — also λ in terms of t.

If a weight W is suspended at the end, and another weight W_1 falls from a height h, giving rise to a velocity v, we have for the common velocity of the bodies after impact, if both are non-elastic, $\text{V} = \frac{\text{W}_1 v}{\text{W}_1 + \text{W}}$, and the *vis viva* of both will be

$$\text{MV}^2 = \frac{\text{W}_1^2 v^2}{g(\text{W}_1 + \text{W})} \quad \text{which equals } \frac{2\text{EK}}{l}\lambda^2, \text{ or twice the work.}$$

$$\therefore \lambda = \frac{\text{W}_1}{\sqrt{(\text{W}_1 + \text{W})}}\sqrt{\frac{hl}{\text{EK}}}.$$

This is only an approximate value, for the inertia of the wire is neglected.

20. VISCOSITY OF SOLIDS.—Experiments show that the principle of equal amplitudes, referred to in the preceding article, is not realized in practice. This is more easily observed in transverse vibrations. The amplitudes grow rapidly less from the first vibration, and the diminution cannot be fully accounted for by the external resistance of air. Professor Thompson of England has shown that there is an internal resistance which opposes motion among the particles of a body, and is similar to that resistance in fluids which opposes the movement of particles among themselves. He therefore called it *viscosity*.* He proved:

1st. That there was a certain internal resistance which he called Viscosity, and which is independent of the elastic properties of metals;

2d. That this force does not affect the co-efficient of elasticity.

The law between molecular friction and viscosity was not discovered.

The viscosity was always much increased at first by the increase of weight, but it gradually decreased, and after a few days became as small as if a lighter weight had been applied. Only one experiment was made to determine the effect of continual vibration; and in that the viscosity was very much increased by daily vibrations for a month.

This latter fact, if firmly established, will prove to be highly important; for it shows that materials which are subjected to constant vibrations, such as the materials of suspension bridges, have within themselves the property of resisting more and more strongly the tendency to elongate from vibration. Experiments will be given hereafter which tend to confirm this fact, when the vibrations are not too frequent or too severe.

But the true viscosity of solids has been fully proved by Mr. Tresca, a French physicist, who showed that when solids are subjected to a very great force, the amount of the force depending upon the nature of the material, that the particles in the immediate vicinity of pressure will *flow* over each other, so as to resemble the flowing of molasses, or tar, or other viscous fluids. Thus, the true viscosity differs entirely in its character from the property recognized by Professor Thompson.

RESISTANCE TO RUPTURE BY TENSION.

21. MODULUS OF STRENGTH.—Many more experiments have been made to determine the ultimate resistance to rupture by tension, than there have to determine the elastic resistance. In the earlier experiments the former was chiefly sought, and more recently all who experimented upon the latter also determined the former.

* Civ. Eng. Jour., vol. 28, p. 322.

The force which is necessary to pull asunder a prismatic bar whose section is one square inch, when acting in the direction of the axis of the bar, is called the *modulus of strength.* This we call T. It expresses the *tenacity* of the material, and is sometimes called the absolute strength and sometimes *modulus of tenacity.*

22. FORMULA FOR THE MODULUS OF STRENGTH; *or the force necessary to break a prismatic bar, when acted upon by a tensile strain.*

Let K=the section of the bar in inches,
T=the modulus of tenacity, and
P=the required force.

It is proved by experiment that the resistance is proportional to the section; hence

$$P=TK \quad . \quad . \quad . \quad . \quad . \quad . \quad . \quad (9)$$

$$\therefore \; T=\frac{P}{K} \quad . \quad . \quad . \quad . \quad . \quad . \quad . \quad (10)$$

From (10) T may be found. In (10) if P is not the ultimate resistance of the bar, then will T be the strain on a unit of section.

From (9) we have

$$K=\frac{P}{T} \quad . \quad . \quad . \quad . \quad . \quad . \quad . \quad (11)$$

which will give the section.

The following are some of the values of T which have been found from experiment by the aid of equation (10).

	Cohesive force or Tenacity in pounds per square inch.
Ash (*English*)	17,000
Oak (*English*)	9,000 to 15,000
Pine (*pitch*)	10,500
Cast Iron *	14,800 to 16,900
Cast Iron (*Weisbach & Overman*)	20,000
Wrought Iron	50,000 to 65,000
Steel wire	100,000 to 120,000
Bessemer steel †	120,000 to 129,000
" " ‡	72,000 to 101,000
Bars of Crucible Steel §	70,000 to 134,000

The most remarkable specimen of cast steel for tenacity which

* Hodgkinson, Bridges. Weale, sup., p. 25.

† Jour. Frank. Inst. Vol. 84, p. 366.

‡ Also experiments by Wm. Fairbairn, Van Nostrand's Ec. En. Mag., Vol. I., p. 273.

§ Do. p. 1009.

is on record, was manufactured in Pittsburgh, Pa. It was tested at the Navy Yard at Washington, D. C., and was found to sustain 242,000 pounds to the square inch! *

For other values see the Appendix.

23. *A vertical prismatic bar is fixed at its upper end, and a weight P_1 is suspended at the other; what must be the upper section, at A, Fig.* **7**, *so as to resist n times all the weight below it, the weight of the bar being considered?*

Let δ = the weight of a unit of volume of the bar, and the other notation as before.

Then $KT = nP_1 + n\delta Kl$.

$$\therefore \quad K = \frac{nP_1}{T - n\delta l} \quad - - - - - - - - - - - - - \quad (12)$$

If $n = 1$, $K = \frac{P_1}{T - \delta l}$; and if $\delta l = T$, $K = \infty$, or no section is possible, and $l = \frac{T}{\delta}$ is the corresponding length of the bar.

24. BAR OF UNIFORM STRENGTH. *Suppose a bar is fixed at its upper extremity, Fig.* **10**, *and a weight P_1 is suspended at its lower extremity; it is required to find the form of the bar so that the horizontal sections shall be proportional to the strains to which they are subjected—the weight of the bar being considered.*

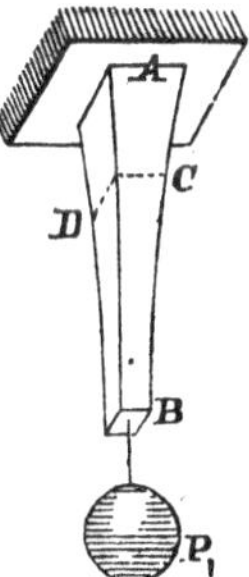

FIG. 10.

Let δ = weight of an unit of volume,

W = weight of the whole bar,

$K_0 = \frac{P_1}{T}$ = the section at B (Eq. (11)),

K_1 = the upper section,

K = variable section, and

x = variable distance from B upwards.

Also let the sections be similar:

Then $P = P_1 + \delta \int K\,dx$ = strain on any section, as D C. But TK is the ability to resist this strain;

$\therefore P_1 + \delta \int K dx = TK$. Differentiate this and we have

$$\delta K\,dx = T dK$$

or $\frac{\delta}{T} dx = \frac{dK}{K}$ which by integrating gives

$$\frac{\delta}{T} x = \text{Nap.} \log K + C \quad - - - - - - \quad (12a)$$

* Am. R. R. Times (Boston), Vol. 20, p. 206.

But for $x = 0$, we have $K = K_0 \therefore C = -$ Nap. log $K_0 = -$ Nap. log $\frac{P_1}{T}$. Hence, Eq. (12a) becomes $\frac{\delta}{T}x =$ Nap. log $\frac{K}{K_0}$ or, passing to exponetials, gives $e^{\frac{\delta}{T}x} = \frac{K}{K_0}$

$$\therefore K = K_0 e^{\frac{\delta}{T}x} = \frac{P_1}{T} e^{\frac{\delta}{T}x} \quad - \quad - \quad - \quad - \quad - \quad - \quad (13)$$

For the upper section $K = K_1$ and $x = l \therefore K_1 = \frac{P_1}{T} e^{\frac{\delta}{T}l}$ - (14)

$$\text{W equals, } \delta \int_0^l K dx = \delta \int_0^l \frac{P_1}{T} e^{\frac{\delta}{T}x} dx = P_1 (e^{\frac{\delta}{T}l} - 1) \quad - \quad - \quad (15)$$

Example. What must be the upper section of a wrought-iron shaft of uniform resistance 1,000 ft. long, so that it will safely sustain its own weight and 75,000 lbs.

Let $T = 10,000$ lbs., and
$\delta = 0.27$ lbs. per cubic inch.
Then Eq. (11) gives $K_0 = 7.5$ sq. inches, and
equation (14) gives $K_1 = 10.37$ inches.

In these formulas *the form of section* does not appear. For tensile strains, the strength is practically independent of the form, but not so for compression. When it yields by crushing, the influence of form is quite perceptible, but not so much so as when it yields by bending under a compressive strain. The latter case will be considered under the head of flexure.

25. STRAINS IN A CLOSED CYLINDER.

Fig. 11.

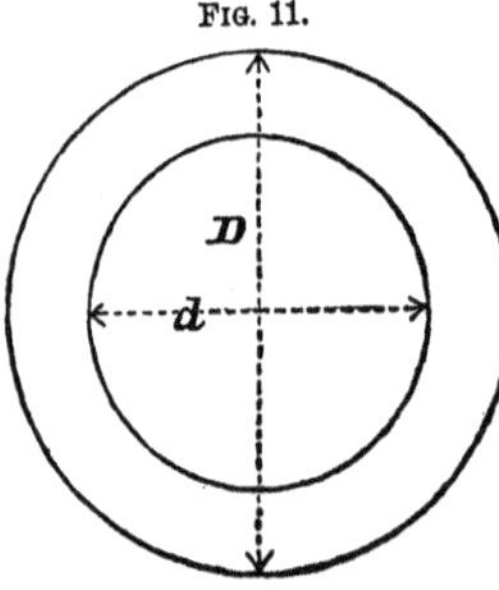

If a closed cylinder is subjected to an *internal pressure*, it will tend to burst it by tearing it open along a rectilineal element, or by forcing the head off from the cylinder, by rupturing it around the cylinder. First, consider the latter case. The force which tends to force the head off is the total pressure upon the head, and the resisting section is the cylindrical annulus.

Let D = the external diameter,
d = the internal diameter,
p = the pressure per square inch, and
t = the thickness of the cylinder.

Then $\frac{1}{4}\pi d^2 p$=the pressure upon the head;
$\frac{1}{4}\pi(D^2-d^2)$=the area of the cylindrical annulus;
$\frac{1}{4}\pi T(D^2-d^2)$=the resistance of the annulus; and,
$2t=D-d$.

Hence, for equilibrium,

$$\tfrac{1}{4}\pi d^2 p=\tfrac{1}{4}\pi T(D^2-d^2)$$

$$\text{or, } d^2 p=2Tt(D+d)=4T(t^2+dt) \quad - \ - \ - \quad (16)$$

which solved gives $t=\left(-1+\sqrt{1+\frac{p}{T}}\right)\frac{d}{2}$ - - - (17)

Next consider the resistance to longitudinal rupturing. As it is equally liable to rupture along any rectilinear element, suppose that the cylinder is divided by any plane which passes through the axis. The normal pressure upon this plane is the force which tends to rupture it, and for a unit of length is

$$pd$$

and the resisting force is

$$2Tt,$$

hence, for equilibrium,

$$2Tt=pd \quad - \ - \ - \ - \ - \ - \ - \ - \quad (18)$$

The value of t from (18) divided by that of t from (16) gives the ratio $\frac{D+d}{d}$, and since D always exceeds d, this ratio is greater than 2; hence there is more than twice the danger of bursting a boiler longitudinally that there is of bursting it around an annulus when the material is equally strong in both directions.

The last equation was established by supposing that all the cylindrical elements resisted equally, but in practice they do not; for, on account of the elasticity of the material, they will be compressed in the direction of the radius, thus enlarging the internal diameter more than the external, and causing a corresponding increase of the tangential stress on the inner over the outer elements. In a thick cylindrical annulus it is necessary to consider this modification.

To find the VARYING LAW OF TANGENTIAL STRAINS, let D and d be the external and internal diameters before pressure, and $D+z$ and $d+y$ the corresponding diameters after pressure. Then, as a first approximation—which is near enough for practice—suppose that the volume of the annulus is not changed, and we have

$$\tfrac{1}{4}\pi(D^2-d^2)=\tfrac{1}{4}\pi(D+z)^2-\tfrac{1}{4}\pi(d+y)^2$$

$$\text{or, } Dz=dy \text{ nearly} \qquad (19)$$

But the strain upon a cylindrical filament varies as its elongation divided by its length; see equation (3). Hence the strain on the external annulus, compared with the internal, is as

$$\frac{\pi(D+z)-\pi D}{\pi D} \text{ to } \frac{\pi(d+y)-\pi d}{\pi d} \text{ or as } \frac{z}{D} \text{ to } \frac{y}{d}$$

which combined with (19) gives

$$\frac{d}{D^2} \text{ to } \frac{1}{d} \text{ or as } d^2 \text{ to } D^2, \text{ or as } r^2 \text{ to } R^2$$

where r and R are radii of the annulus.

Hence, *the strain varies inversely as the square of the distance from the axis of the cylinder.*

To FIND THE TOTAL RESISTANCE, let

x = the variable distance from the axis of the cylinder,
T = the modulus of rupture, or of strain, and
t = the thickness of the annulus.

Then Tdx is the strain on an element at a distance r from the axis of the cylinder, or otherwise upon the inner surface of the cylinder; and according to the principle above stated, $T\frac{r^2}{x^2}dx$ is the strain on any element, and the total strain on both sides is

$$2Tr^2\int_r^R\frac{dx}{x^2}=2T\frac{rt}{r+t} \qquad (20)$$

If $t = r$, this becomes

$$Tt$$

which compared with equation (18) shows that when the thickness equals the radius, the resistance is only half what it would be if the material were non-elastic. In (20) if t is small compared with r, it becomes $2Tt$ nearly, which is the same as equation (18).

If the ends of the cylinder are capped with hemispheres, the stress upon an elementary annulus at the inner surface is $2\pi Trdx$.* Proceeding as before, and we find that the total stress necessary to force the hemispherical heads off is

$$2\pi T\frac{r^2t}{r+t} \quad - \; - \; - \; - \; - \; - \; - \; - \quad (21)$$

which is also the stress necessary to force asunder a sphere by internal pressure, when the elasticity is considered.

If cylinders are formed by riveting together plates of iron, their strength will be much impaired along the riveted section. The condition of the riveted joint will doubtless have much more to do with the strength than the compressibility of the material, and will hereafter be considered.

---

* T. J. Rodman says *the resistance on any elementary annulus is* $T2\pi xdx$ (Exp. on metal for cannon, p. 44); but it appears to me that, to make his expression correct, T must be the modulus at any element considered, and hence variable, whereas it should be constant. The strain on any elementary annulus whose distance is x from the centre of the sphere, is $T2\pi xdx, \frac{r^2}{x^2} = 2\pi r^3T\frac{dx}{x^2}$; and the total resistance is the integral of this expression between the limits of r and $r+t$.

26. RESISTANCE OF GLASS GLOBES TO INTERNAL PRESSURE.

EXPERIMENTS OF WM. FAIRBAIRN.

Description of the glass.	Diameter in inches.	Thickness in inches.	Bursting pressure in lbs. per square inch.	Bursting pressure in lbs. per square inch of section.
Flint-glass...........	4.0 × 3.98	0.024	84	3504
	4.0 × 3.98	0.025	93	3720
	4	0.038	150	3947
	4.5 × 4.55	0.056	280	5625
	6	0.059	152	3864
			Mean..........	4132

Description of the glass.	Diameter in inches.	Thickness in inches.	Bursting pressure in lbs. per square inch.	Bursting pressure in lbs. per square inch of section.
Green-glass..........	4.95 × 5.0	0.022	90	5113
	4.95 × 5.0	0.020	85	5312
	4.0 × 4.05	0.018	84	4666
	4.0 × 4.03	0.016	82	5126
			Mean..........	5054

Description of the glass.	Diameter in inches.	Thickness in inches.	Bursting pressure in lbs. per square inch.	Bursting pressure in lbs. per square inch of section.
Crown-glass..........	4.2 × 4.35	0.025	120	5040
	4.05 × 4.2	0.021	126	6000
	5.9 × 5.8	0.016	69	6350
	6.0 × 6.3	0.020	86	6450
			Mean..........	5960

The following table exhibits the tensile strength of cylindrical glass bars according to the experiments of Fairbairn:—

Description of the glass.	Area of specimen in inches.	Breaking weight in lbs.	Tenacity per square inch.
Annealed flint-glass...	0.255	583	2286
	0.196	254	2540
Green-glass...........	0.220	639	2896
Crown-glass...........	0.229	583	2546

As might have been anticipated, the tenacity of bars is much less than globes; for it is difficult to make a longitudinal strain without causing a transverse strain, and the latter would have a very serious effect; it is also probable that the outer portion of the annealed glass is stronger than the inner, and there is a larger amount of surface compared with the section, in globes than in cylinders.

RIVETED PLATES.

27. RIVETED PLATES are used in the construction of boilers, roofs, bridges, ships, and other frames. It is desirable to know the best conditions for riveting, and the strength of riveted plates compared with the solid section of the same plates. The common way of riveting is to *punch* holes through both plates, into which red-hot bolts or rivets are placed, and headed down while hot. The process of punching strains, and hence weakens, the material. A better way is to *bore* the holes in the plates, and then rivet as before. The holes in the separate plates should be exactly opposite each other, so that there will be no side strain on the plates caused by driving the rivets home, and to secure the best effects of the rivets themselves. They are sometimes placed in single and sometimes in double rows,

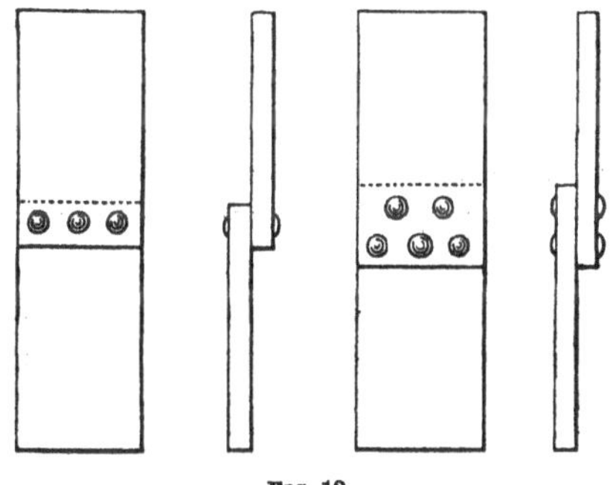

FIG. 12.

and experiment shows that the latter possesses great advantage over the former. Experiments have been made upon plates of the form shown in Fig. 12, both with lap and butt-joints, and with single and double rows of rivets.*

* Lond. Phil. Transactions, part 2d, 1850, p. 677.

Table showing the strength of single and double riveted plates.

Cohesive strength of the plates in lbs. per square inch. T.	Strength of single-riveted joints of equal section to the plates, taken through the line of rivets. Breaking weight in lbs. per square inch.	Strength of double-riveted joints of equal section to the plates, taken through the line of rivets. Breaking weight in lbs. per square inch.
57,724	45,743	52,352
61,579	36,606	48,821
58,322	43,141	58,286
50,983	43,515	54,594
51,130	40,249	53,879
49,281	44,715	53,879
Mean. .54,836	42,328	53,635

It will be observed that in double-riveting there is but little loss of strength, while there is considerable loss in single-riveting. In the preceding experiments the solid section of the plates, taken through the centre of the rivet-holes, was used; but, as Fairbairn justly remarks, we must deduct 30 per cent. for metal actually punched out to receive the rivets. But as only a few rivets came within the limits of the experiments, and as an extensive combination of rivets must resist more effectually, and as something will be gained by the friction between the plates, it seems evident that we may use more than 60 per cent. of the strength of riveted plates as indicated above. Fairbairn says we may use the following proportions:—

Strength of plates..............................	100
Strength of double-riveted plates................	70
Strength of single-riveted plates.................	56

Size and distribution of rivets.—The best size of the rivets, the distance between them, and the proper amount of lap of the plates, can be determined only by long experience, aided by experiments. Fairbairn gives the following table as the results of his information upon this important subject, to make the joint steam or water tight:—

Table showing the strongest forms and best proportions of riveted joints, as deduced from experiments and actual practice. (*Useful Information for Engineers, 1st Series, p.* 285.)

Thickness of plates in inches. t.	Diameter of the rivets in inches. d.	Length of rivets from the head in inches. l.	Distance of rivets from centre to centre in inches. a.	Quantity of lap in single joints in inches. b.	Quantity of lap in double-riveted joints in inches. c.
$\frac{3}{16}$ to $\frac{4}{16}$	2 t	$4\frac{1}{2}$ t	6 t	6 t	10 t
$\frac{5}{16}$	"	"	5 t	"	"
$\frac{6}{16}$	"	"	"	$5\frac{1}{2}$ t	$8\frac{1}{4}$ t
$\frac{8}{16}$ to $\frac{12}{16}$	$1\frac{1}{2}$ t	"	4 t	$4\frac{1}{2}$ t	$6\frac{3}{4}$ t

28. STRENGTH OF IRON IN DIFFERENT DIRECTIONS OF THE ROLLED SHEET.*

In obtaining specimens for these experiments, care was generally taken to have them cut in different directions of the rolling, longitudinally and transversely, and in some cases *diagonally*, to that direction. The table will be found to indicate the direction of slitting in each case, and the comparison contained in the table is given to show what information the inquiry has elicited.

The comparison is made principally on the *minimum* strength of each bar, being that which can alone be relied on in practice; for if the strength of the weakest point in a boiler be overcome, it is obviously unimportant to know that other parts had greater strength. In one case, however, two bars, one cut across the direction of rolling, and the other longitudinally, were, after being reduced to uniform size, broken up cold, with a view to this question. The result showed that the length-strip was $7\frac{1}{10}$ per cent. stronger than the one cut crosswise, considering the tenacity of the latter equal to 100. Of the other sets, embracing about 40 strips cut in each direction, it appears that some kinds of boiler iron manifest much greater inequality in the two directions than others. It is in certain cases not much over one per cent., and in others exceeds twenty, and as a mean of the whole series it may be stated to amount to six per cent. of the strength of the cross-cut bars. The number of trials on those cut diagonally is not perhaps sufficiently great to warrant a general deduction; but, so far as they go, they certainly indicate that the strength in this direction is less than either of the others.

* Experiments of Franklin Institute.

Had we compared the mean instead of the least strength of bars as given in the table, the result would not have differed materially in regard to the relative strength in the respective directions.

The boiler-iron manufactured by Messrs. E. H. & P. Ellicott, which was tried in all these modes of preparation of specimens, gave the following results:

1. When tried at *original sections*, seven experiments on length-sheet specimens gave a mean strength of 55285 lbs. per square inch, the lowest being 44399 lbs., and the highest 59307 lbs. Fourteen experiments on cross-sheet specimens gave a mean of 53896 lbs., the lowest result being 50212 lbs., the highest 58839 lbs.; and six experiments on strips cut diagonally from the sheet exhibited a strength of 53850 lbs., of which the lowest was 51134 lbs., and the highest 58773 lbs.

2. When tried by filing notches on the edges of the strips, to remove the weakening effect of the shears, the *length-sheet* bars gave, at fourteen fractures, a mean strength of 63946 lbs., varying between 56346 lbs. and 78000 lbs. per square inch. The cross-sheet specimens tried after this mode of preparation exhibited, at three trials, a mean strength of 60236 lbs., varying from 55222 lbs. to 65143 lbs.; and the *diagonal* strips, at four trials, gave a mean result of 53925 lbs., the greatest difference being between 51428 lbs. and 56632 lbs. per square inch.

3. Of strips reduced to uniform size by filing, four comparable experiments on those cut lengthwise of the sheet gave a mean strength of 63947 lbs., of which the highest was 67378 lbs., and the lowest 60594 lbs.

Cross-sheet specimens, tried after the same preparation, exhibited, at thirty-three fractures, a mean of 50176 lbs., of which the highest was 65785 lbs., and the lowest 52778 lbs. No bar cut diagonally was reduced to uniform size.

From the foregoing statements it appears that by filing in notches and filing to uniformity, we obtain results 63946 lbs. and 63947 lbs. for the strength of strips cut lengthwise, differing from each other by only a single pound to the square inch, and that by these two modes of preparation the cross-sheet specimens gave respectively 60236 lbs. and 60176 lbs., differing by only 60 lbs. to the square inch. This seems to prove that by both methods of preparing the specimens the accidental weakening effect of slitting had been removed by separating all that portion of the metal on which it had been exerted. Hence we may infer that the differences between length-sheet and cross-sheet specimens are really and truly ascribable to a difference of texture in the two directions, which will be seen to amount, in the case of filing in notches, to 6.15 per cent., and in that of filing to uniformity, to 6.26 per cent. of strength of cross-sheet specimens.

Table of the comparative view of the strength of specimens of ten different sorts of boiler and one of bar iron, in the longitudinal, transverse, and diagonal direction of the rolling, as deduced from the least strength of each specimen, and the average minimum of each sort of iron, in each direction in which it was tried.

No. of specimen referred to.	Strength in the longitudinal direction.	Strength in the transverse direction.	No. of specimen referred to.	Strength in the longitudinal direction.	Strength in the transverse direction.	Strength in the diagonal direction.
2	58977		125	57182	Tilted.	
3	53828		130	Tilted.	57789	
4	47167		133	do.	53176	
6		52280	135	do.		47738
8		50103	137	do.		50358
Mean	53324	51191	Mean	57182	55882	49048
42	51653	Puddled.	142	44399		
43	44102	do.	143	53135		
44	53836	do.	146	60594		
46	59262	H'd pla.*	148		52468	
48	59418	do.	149		52228	
49	57565	do.	150		56869	
51	H'd pl.	59656	151		53811	
53	H'd pla.	56062	152		56073	
56	Puddled.	57926	154			51134
58	do.	50570	157			52102
59	48308	Puddled.	160		53862	
60	58684	do.	162		50212	
61	52869	do.	164	56346		
62	57612	do.	167	56682		
64	Puddled.	45392	169	54361		
65	do.	51255	171			55612
68	57929	H'd pla.	174			51425
70	47638	do.				
71	H'd pla.	54634	Mean	54253	53646	52568
73	do.	52657				
74	do.	49351				
Mean	54074	53049	226		49053	
			227		53699	
			228		40643	
			229		46473	
			230	49368		
			Mean	49368	47467	

* Hammered and rolled into plates.

The specimens from 42 to 74 were partly puddled iron, and partly Juniata blooms, hammered and rolled into plate. The length and cross-sheet specimens of these two kinds must be compared separately.

All the experiments on No. 228 (cross) and 230 (length) were made at ordinary temperatures with a view to this comparison.

29. TENSILE STRENGTH OF WROUGHT IRON AT VARIOUS TEMPERATURES.

Mr. Fairbairn has made experiments upon rolled plates of iron, and rods or rivet iron, at various temperatures. The former were broken in the direction of the fibre and across it. The specimen when subjected to experiment was surrounded with a vessel into which freezing mixtures were placed to produce the lower temperatures, and oil heated by a fire underneath to produce the high temperatures. The experiments were made upon Staffordshire plates, which are inferior to several other kinds in common use. The following table gives a summary of the results:—

Table showing the Resistance of Staffordshire Plates at Different Temperatures.

No. of experiment.	Temp. Fahr.	Section of plate in inches.	Breaking weight in lbs.	Breaking weight per square inch, lbs.	Mean breaking weight per square inch, lbs.	Remarks.
1	0°	0.6868	33,660	49,009	49,009	With.
2	60	0.7825	31,980	40,357		Across.
3	60	0.6400	27,780	43,406	44,498	Across.
4	60	0.6368	31,980	50,219		With.
5	110	0.6633	29,460	44,160		Across.*
6	112	0.6800	28,620	42,088	42,291	With.
7	120	0.8128	37,020	40,625		With.
8	212	0.8008	31,980	39,935		With.
9	212	0.6633	30,300	45,680	45,005	Across.
10	212	0.6800	33,660	49,500		With.
11	270	0.6432	28,620	44,020	44,020	With.
12	340	0.6400	31,980	49,968	46,018	With.†
13	340	0.6800	28,620	42,088		Across.
14	395	0.6666	30,720	46,086	46,086	With.
15	Scarcely red	0.6200	23,520	38,032	34,272	Across.
16	Dull red	0.6076	18,540	30,513		Across.‡

* Too high; fracture very uneven.

† Too low; tore through the eye.

‡ Too high; the specimen broke with the first strain.

The mean values given in the sixth column of this Table exhibit a remarkable degree of uniformity in strength for all temperatures, from 60 degrees to 395 degrees. The single example at 0 degrees gives a higher value than the mean of the others, but not higher than for some of the specimens at higher temperatures. At red heat the iron is very much weakened. This fact should be noticed in determining the strength of boiler-flues, as they are often subjected to intense heat when not covered with water.

The experiments upon rivet iron were made with the same machine, and in the same manner, the results of which are shown in the following table:—

Table showing the Results of Experiments on Rivet Iron at Different Temperatures.

No. of experiment.	Temp. Fahr.	Section, inches.	Breaking weight in lbs.	Breaking weight per square inch, lbs.	Mean breaking weight per square inch, lbs.	Remarks.
17	−30°	[illegible]	[illegible]	[illegible]	[illegible]	Too low.
18	+60	0.2485	15,400	61,971	62,816	Too low.
19	60	......	15,820	63,661		Too low.
20	114	......	17,605	70,845	70,845	
21	212	......	20,545	82,676	79,271	
22	212	0.1963	14,560	74,153		
23	212	0.2485	20,125	80,985		
24	250	0.1963	16,135	82,174	82,636	
25	270	0.2485	20,650	83,098		
26	310	0.1963	15,820	80,570	84,046	
27	325	0.1963	17,185	87,522		
28	415	0.2485	20,335	81,830	83,943	
29	435	......	21,385	86,056		
30	Red heat.	......	8,965	36,076	35,000	Too high.

From this Table we see that there is a gradual increase of strength from 60 degrees to 325, where it appears to attain its maximum. The increase is a very important amount, being about 30 per cent.

It is a little remarkable that the specimen at minus 30 degrees is stronger than the mean of the two at 60 degrees; but we observe, as before, that it is not as strong as some of the single specimens at higher degrees.

Mr. Johnson, when in the employ of the Navy Department, in 1844, made some experiments to determine the effects of thermo-tension upon different kinds of iron.* He took two bars of the same kind of iron, and of the same size, and broke one while cold. He then subjected the other to the same tension when heated 400 degrees, after which the strain was relieved, and the bar was allowed to cool, and the permanent elongation noted, after which it was broken by an additional load. It will thus be seen that the experiments were not conducted in the same way as those by Fairbairn. The following table gives the results of his experiments:

The Results of Experiments on Thermo-Tension, at 400° Temperature.

KIND OF IRON.	Strength cold.	Strength after heated with thermo-tension.	Section.	Gain of length under thermo-tension with a strain equal to the strength when cold.	Gain of strength.	Total gain relative to the weight.
	Tons.	Tons.	Inches.	Per cent.	Per cent.	Per cent.
Tredegar, round...	60	71.4	1.91	6.51	19.00	25.51
Tredegar, round...	60	72.0	1.91	(6.51)	20.00	26.51
Tredegar, square bar	60	67.2	1.69	6.77	12.00	18.77
Tredegar, r'nd, No. 3	58	68.4	1.15	5.263	17.93	23.19
Salisbury, round ...	105.87	121.0	3.59	3.73	14.64	18.37
Mean....................				5.75	16.64	22.40

Remarks.—From the two former sets of experiments, pp. 36 and 37, it appears that the strength of the iron was increased by an increase of temperature at the time the bar was broken, and by the latter that it was not only increased, but, by being subjected to severe tension while at a high temperature, the increased strength was not lost by cooling. It hardly seems probable that this increased strength would be retained indefinitely, and hence it is important to know how long it was after the piece was cooled before it was broken.

These results are confirmed by the experiments of the committee of the Franklin Institute, as shown by the following table. See Journal of the Franklin Institute, vol. xx., 3d series, p. 22.

---

* Senate Doc., No. 1, 28th Cong., 2d Sess., 1844–5, p. 639.

ABSTRACT OF A TABLE

Of the comparative view of the influence of high temperatures on the strength of iron, as exhibited by 73 experiments on 47 different specimens of that metal, at 46 different temperatures, from 212° to 1317° Fahr., compared with the strength of each bar when tried at ordinary temperatures, the number of experiments at the latter being 163.

No. of the experiment.	Temperature observed at moment of fracture.	Strength at ordinary temperature.	Strength at the temperature observed.
1	212°	56736	67939
2	214	53176	61161
3	394	68356	71896
9	440	49782	59085
10	520	54934	58451
15	554	54372	61680
20	568	67211	76763
25	574	76071	65387
40	722	57133	54441
45	824	59219	55892
50	1037	58992	37764
58	1245	54758	20703
59	1317	54758	18913

Remark.—According to these experiments, as shown in the 4th column, the strength increases with the temperature to 394 degrees, when it attains its maximum; although in some cases the strength was increased by increasing the temperature to 568 degrees. By comparing the 3d and 4th columns we see that the strength is greater for all degrees from 212° to 574° than it is at ordinary temperatures, but above 574° it is weaker. The experiments on Salisbury iron showed that the maximum tenacity was 15.17 per cent. greater than their mean strength when tried cold. The committee above referred to determined the maximum strength of about half the specimens used in the preceding table by actual experiment, and calculated it for the others; and from the results derived the following empirical formula for the diminution in strength below the maximum for high degrees of heat:—

$$D^5 = c(\theta - 80)^{13}$$

in which D is the diminution after it has passed the maximum,
θ the temperature Fahrenheit, and
c a constant.

This formula appears to be sufficiently exact for all temperatures between 520° and 1317°.

30. EFFECT OF SEVERE STRAINS UPON THE ULTIMATE TENACITY OF IRON RODS.—Thomas Loyd, Esq., of England, took 20 pieces of $1\frac{3}{8}$ S. C. bar iron, each 10 feet long, which were cut from the middle of as many rods. Each piece was cut into two parts of 5 feet each, and marked with the same letter. A, B, C, &c., were first broken, so as to get the average breaking strain. A2, B2, &c., were subjected to the constant action of three-fourths the breaking weight, previously found, for five minutes. The load was then removed, and the rods afterwards broken.

*Results of the Experiments.**

FIRST.		SECOND.	
Mark on the bars.	Breaking weight in tons (gross).	Mark.	Breaking weight in tons.
A	33.75	A 2	33.75
B	30.00	B 2	33.00
C	33.25	C 2	33.25
D	32.75	D 2	32.25
E	32.50	E 2	32.50
F	33.25	F 2	33.00
G	32.75	G 2	33.00
H	33.25	H 2	33.50
I	33.50	I 2	32.75
J	33.50	J 2	33.25
K	32.25	K 2	32.50
L	32.25	L 2	31.50
M	30.25	M 2	32.75
N	34.25	N 2	34.00
O	31.75	O 2	32.50
P	29.75	P 2	31.00
Q	33.50	Q 2	33.75
R	33.75	R 2	33.75
S	33.00	S 2	33.25
T	32.25	T 2	31.00
Mean..........	32.57		32.81

We here see that a strain of 25 tons, or three-fourths the breaking weight, did not weaken the bar.

* Fairbairn, Useful Information for Engineers, First Series, p. 313.

These experiments indicate that a frame or bridge may be subjected to a severe strain of three-fourths of its strength for a short time without endangering its ultimate strength.

31. EFFECT OF REPEATED RUPTURE.—The following experiments were made at Woolwich Dockyard, England. The same bar was subjected to three or four successive ruptures by tensile strains. They show the remarkable fact, that while great strains impair the elasticity, as shown by Hodgkinson, yet they do not appear to diminish the ultimate tenacity. This fact is important, for it shows that iron, which has been broken by tension in a structure, may safely be used again for any strain less than that for which it was broken.

Table showing the effect of repeated Fracture on Iron Bars.

Mark.	First breakage.		Second breakage		Third breakage.		Fourth breakage.		Reduced from sectional area of 1.37 sqr. inches to
	Tons.	Stretch in 54 inches.	Tons.	Stretch in 36 inches.	Tons.	Stretch in 24 inches.	Tons.	Stretch in 15 inches.	
		In.		In.		In.			
A	33.75	0.9125	35.50	0.200					
B	33.75	0.9250	35.25	0.225	37.00	1.00	38.75		1.25
E	32.50	0.9250	34.75	0.125					
F	33.25	1.0500	35.50	0.112	37.25	0.62	40.40		1.18
G	32.75	0.8500	35.00	0.125	37.50		40.41		1.25
H	33.75	1.0625	36.25	0.187					
I	33.50	0.8375	34.50	0.62	36.50	1.50			
J	33.50	0.9250	36.00	0.025	36.75	1.12	41.75		1.25
L	32.25	Defect'e	36.50	0.150	37.75		41.00	0.31	1.25
M	30.25	Defect'e	36.50	.62	37.75	0.60	38.50	0.06	1.25
Mean..	32.95		35.57		37.21		40.16		1.24
Mean pr. sq. in.	24.04		25.93		27.06		29.20		0.90

We thus see that while the section is reduced 10 *per cent., the strength is apparently increased over* 20 *per cent.* It is not, however, safe to infer that the strength is *actually* increased, for it is probable that it broke the first time at the weakest point, and the next time at the next weakest point, and so on.

We also observe that the total elongations are not proportional to the tensile strains, which is in accordance with the results of other experiments.

ANNEALING.

32. ANNEALING is a process of treating metals so as to make them more ductile. To secure this, the metals are subjected to a high heat and then allowed to cool slowly. *Steel* is softened in this way, so that it may be more easily worked. Campin * says that *steel* should not be overheated for this purpose. Some bury the heated steel in lime; some in cast-iron borings; and some in saw-dust. He (Campin) says the best plan is to put the steel into an iron box made for the purpose, and fill it with dust-charcoal, and plug the ends up to keep the air from the steel; then put the box and its contents into a fire until it is heated thoroughly through, and the steel to a low red heat. It is then removed from the fire, and the steel left in the box until it is cold. Tools made of annealed steel will, in some cases, last much longer than those made of unannealed steel.

But it appears from the following table that it weakens *iron* to anneal it.

Table of the strength of Wrought Iron Annealed at Different Temperatures.

No. of comparisons.	Strength at ordinary temp. before annealing.	Temperature at which annealing took place.	Strength at the annealing temperature.	Strength after annealing and cooling.	Ratio of diminution of strength.
1	57,133	1037°	37,764	55,678	0.025
5	53,774	1155	21,967	45,597	0.152
10	52,040	1245	20,703	38,843	.253
15	48,407	Bright welding heat.	......	38,676	.201
17	73,880	Low welding heat.	......	53,578	.275
18	76,986	Bright welding heat.	......	50,074	.349
19	89,162	Low welding heat.	......	48,144	.460

33. THE STRENGTH OF IRON AND STEEL ALSO DEPENDS largely upon the processes of their manufacture, and their treatment afterwards. The strength of wrought iron depends upon the ore of which it is made; the manner in which it is smelted and puddled, the temperature at which it is hammered, and the amount of hammering which it receives in bringing it into shape. The same remark applies to cast steel. If the former is hammered when it is comparatively cold, it will weaken it, especially if the blows are heavy; but the latter, steel, may be greatly damaged, or even rendered worthless by excessive heat, and it is greatly improved by hammering when comparatively cold. For the effect of tempering on the crushing strength, see Article 50.

* Campin's "Practical Mechanics," p. 364.

34. PROLONGED FUSION OF CAST IRON.—Cast iron is also subjected to great modifications of strength on account of the manipulations to which it is or may be subjected in its manufacture and preparations for use. The strength in some cases is greatly increased by keeping the metal in a fused state some time before it is cast. Major Wade made experiments upon several kinds of iron, all of which were increased in strength with prolonged fusion (see Rep. p. 44), one example of which is given in the following

Table showing the Effects of Prolonged Fusion.

		Tensile Strength in lbs. per sq. in.
Iron in fusion	½ hour	17,843
" " "	1 "	20,127
" " "	1½ hour	24,387
" " "	2 hours	34,496

35. EFFECT OF REMELTING CAST IRON.—But the greatest effect was produced by remelting. The density, tenacity, and transverse strength were all increased by it, within certain limits. For instance, a specimen of No. 1 Greenwood pig-iron gave the following results :—(Rep., p. 279.)

Table showing the Effects of Remelting.

No. 1 Greenwood iron.	Density.	Tensile Strength.
Crude pig-iron	7.032	14,000
" remelted once	7.086	20,900
" " twice	7.198	30,229
" " three times	7.301	35,786

But there is a point beyond which remeltings will weaken the iron. Mr. Fairbairn made an experiment in which the strength of the iron was increased for twelve remeltings, and then the strength decreased to the eighteenth, where the experiment terminated. In some cases no improvement is made by remelting, but the iron is really weakened by the process; so that it becomes necessary to determine the character of each iron under the various conditions by actual experiment.

The laws which govern Greenwood iron were so thoroughly

determined that the results which will follow from any given course of treatment may be predicted with much certainty. (Rep., p. 245.)

By mixing grades Nos. 1, 2, and 3, and subjecting them to a third fusion, *one* specimen was obtained whose density was 7.304, and whose tenacity was 45,970 pounds, which is the strongest specimen of cast iron ever tested. (Rep., p. 279.)

As a general result of these experiments, Major Wade remarks (p. 243), "that the softest kinds of iron will endure a greater number of meltings with advantage than the higher grades. It appears that when iron is in its best condition for casting into proof bars (that is, small bars for testing the metal) of small bulk, it is then in a state which requires an additional fusion to bring it up to its best condition for casting into the massive bulk of cannon."

36. THE MANNER OF COOLING also affects the strength. It was found that the tensile strength of large masses was increased by slow cooling; while that of small pieces was increased by rapid cooling. (Rep., p. 45.)

37. THE MODULUS OF STRENGTH IS MODIFIED, we thus see, by a great variety of circumstances; and hence it is impossible to assign any *arbitrary* value to it for any material, that will be both safe and economical; but its value must be determined, in any particular case, by direct experiment, or something in regard to the quality of the material must be known before its approximate value can be assumed.

38. SAFE LIMIT OF LOADING.—Structures should not be strained so severely as to damage their elasticity. According to Article 9, it appears that a weight suddenly applied will produce twice the elongation that it will if applied gradually or by increments. Hence, structures which are subjected to shocks by sudden applications of the load, should be so proportioned as to resist more than double the load as a constant dead-weight without straining it beyond the elastic limit.

This method of indicating the limits, suggested by M. Poncelet, is perfectly rational; but, unfortunately, the elastic limits have not been as closely observed and as thoroughly determined by experimenters as the limit of rupture. The latter was formerly considered more important, and hence furnished the basis for determining the safe limit of the load. Observations on good constructions have led engineers to adopt the following values as mean results for permanent strains on bars:—

For wood,	$\frac{1}{10}$	The load which would produce rupture.
For wrought iron,	$\frac{1}{3}$ to $\frac{1}{5}$	
For cast iron,	$\frac{1}{4}$ to $\frac{1}{6}$	

Further observations will be made upon this subject in the latter part of this volume.

CHAPTER II.

COMPRESSION.

38. RESISTANCE TO COMPRESSION also divides itself into two general problems—*elastic* and *ultimate*. The law of elastic resistance for compression may be as readily found as that for tensive resistance; but the law of resistance to crushing is very complex. If the pieces which are subjected to this stress are long, they will bend under a heavy stress, unless they are confined, and when they bend they break partly by bending and partly by crushing. If the pieces are very short, compared with their diameter, they may be crushed without being bent; but even in this case, with granular substances, the yielding is more or less peculiar, dividing off in pieces at certain angles with the line of pressure. The results of some experiments will now be given, which will enable us to test the prevailing theories upon this subject.

ELASTIC RESISTANCE.

TABLE

Showing the compression, permanent set, and coefficient of elasticity * *of a solid cylinder* 10 *inches long and* 1.382 *inch diameter.*

Weight per square inch of section in lbs.	Compression per inch of length.	Permanent set per inch of length.	Coefficient of elasticity.
1,000	0.000090	0.	11,111,000
2,000	0.000170	0.	11,824,000
3,000	0.000255	0.000005	11,843,100
4,000	0.000320	0.000015	12,500,000
5,000	0.000385	0.000025	12,987,000
6,000	0.000455	0.000030	13,189,000
7,000	0.000505	0.000035	13,861,300
8,000	0.000575	0.000045	13,813,000
9,000	0.000645	0.000055	13,952,000
10,000	0.000705	0.000070	14,196,000
15,000	0.001035	0.000170	14,492,000
20,000	0.001395	0.000300	14,337,000
25,000	0.001825	0.000495	13,687,900
30,000	0.002380	0.000820	12,602,300

* The author computed the coefficients of elasticity from the other data of the table.

39. COMPRESSION OF CAST-IRON.—Captain T. J. Rodman, in his report upon metals for cannon, page 163, has given the results of experiments upon a piece of cast-iron, which was taken from the body of the same gun as was the specimen referred to on page 11 of this work, the results of which are given on the preceding page.

We observe that the coefficient of elasticity is much less for the first strains than for those that follow. It thus appears that this metal resists more strenuously after it has been somewhat compressed than at first. The coefficient of elasticity is considerably less than for the corresponding piece, as given on page 11. The difference is very much greater than that found by Mr. Hodgkinson in the specimens which he used in his experiments. He took bars 10 feet long, and about an inch square, and fitted them nicely in a groove so that they could not bend, and occasionally, during the experiment, they were slightly tapped to avoid adherence. The metal was the same kind as that used in the experiment recorded on page 13.

TABLE

Giving the results of experiments by Mr. Hodgkinson on bars of cast-iron 10 *feet long.*

Pressure per square inch of section. P.	Compression per inch of length.		Coefficient of elasticity per square inch.	Error in parts of P of the formula $P=170,763\lambda_0 -36,318\lambda_0^2$
	Total. λ_0	Permanent.		
lbs.	in.	in.	lbs.	
2064.74	0.0001561	0.00000391	13,231,300	$-\frac{1}{36}$
4129.49	0.0003240	0.00001882	12,764,910	$-\frac{1}{816}$
6194.24	0.0004981	0.00003331	12,442,300	$+\frac{1}{52}$
8258.98	0.0006565	0.00005371	12,585,100	$+\frac{1}{1330}$
10323.73	0.00082866	0.00007053	12,467,100	$+\frac{1}{312}$
12388.48	0.00100250	0.00009053	12,357,200	$+\frac{1}{248}$
14453.22	0.00128025	0.00011700	12,253,700	$+\frac{1}{179}$
16517.97	0.00136150	0.00014258	12,141,200	$+\frac{1}{139}$
18582.71	0.00154218	0.00017085	12,058,100	$+\frac{1}{161}$
20647.46	0.00171866	0.00020685	12,021,800	$+\frac{1}{629}$
24776.95	0.00208016	0.00036810	11,920,000	$-\frac{1}{176}$
28906.45	0.00247491	0.00045815	11,687,400	$-\frac{1}{275}$
33030.80	0.0029450	0.00050768	11,222,750	$+\frac{1}{64}$
37159.65	0.003429			

In this case the highest coefficient of elasticity results from the smallest strain which is recorded. The difference in this respect between this example and the preceding one results doubtless from the internal structure of the iron. The coefficient in both these cases is much less than that found for other kinds of cast-iron, as is shown in the table of resistances in the Appendix.

Mr. Hodgkinson proposed the empirical formula,—$P = 170{,}763\lambda_c - 36{,}318\lambda_c^2$,—to represent the results of the experiments; and although it may represent more nearly the results of a greater range of strains than equation (3), yet there is no advantage in its use in practice.

40. COMPRESSION OF WROUGHT IRON.

Mr. Hodgkinson also made experiments upon bars of wrought iron in precisely the same manner as upon those of cast iron, the results of which are given in the following

TABLE

*Giving the results of experiments by Mr. E. Hodgkinson on bars of wrought iron, each of which was ten feet long.**

Weight producing the compression.	1st Bar. Section=1.025 × 1.025 sq. in.		2d Bar. Section=1.016 × 1.02 sq. in.	
	Amount of Compression.	Value of E.	Amount of Compression.	Value of E.
lbs.	inch.	lbs.	inch.	lbs.
5098	0.028	20,796,500	0.027	21,864,000
9578	0.052	21,049,000	0.047	23,595,000
14058	0.073	21,979,000	0.067	24,273,000
16298	0.085	21,343,000		
18538	0.096	22,156,000	0.089	24,108,000
20778	0.107	22,160,000	0.100	24,038,000
23018	0.119	23,587,000	0.113	23,587,000
25258	0.130	22,095,000	0.128	23,679,000
27498	0.142	22,111,000	0.143	22,259,000
29738	0.152	21,938,000	0.163	21,139,000
31978	0.174	20,979,000	0.190	19,478,000
In ½ hour.			0.261	
Again after ½ hour.			0.269	
Then repeated.			0.328	

* The coefficients of elasticity were computed by the author.

42. GRAPHICAL REPRESENTATION.—These two cases are graphically represented in Fig. 13. It is seen from the tables that the compressions are quite uniform for a large range of strains, and hence equation (2), page 17, is applicable to compressive strains when within the elastic limits. In the case of the wrought-iron bars, the first one attains its maximum coefficient of elasticity for a strain somewhat less than one-half its ultimate resistance to crushing, and the second bar at about one-third its ultimate resistance.

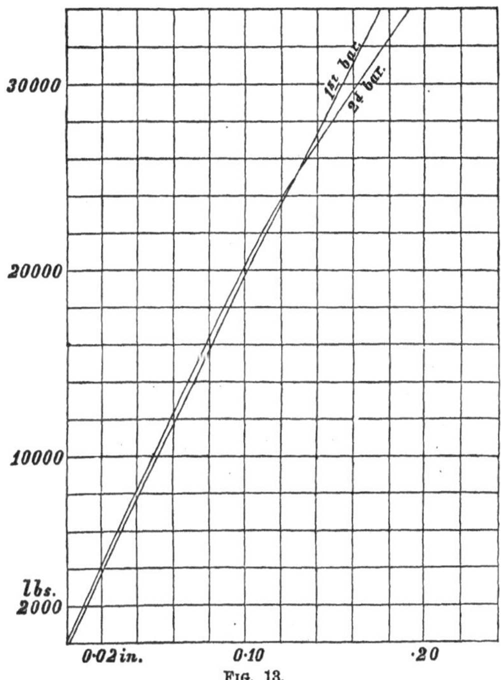

FIG. 13.

43. COMPARATIVE RESISTANCE OF CAST AND WROUGHT IRON.—The coefficient of elasticity is a measure of the compressibility of metals. Hence, an examination of the two preceding tables shows that of the specimens used in these experiments, the cast iron was compressed nearly twice as much as the wrought iron for the same strains. An examination of the table of resistances, in the appendix, shows that for a mean value wrought iron is compressed about two-thirds as much as cast iron for the same strain. The same ratio evidently holds for tension. This is contrary to the popular notion, that cast iron is *stiffer* than wrought iron; for it follows from the above that a cast-iron bar may be stretched more, compressed more, and bent more, than an equal wrought iron one with the same force under the same circumstance, and in some cases, the changes will be twice as great. One reason why cast is considered stiffer than wrought iron probably is, that wrought iron does not fail suddenly as a general thing, but it can be seen to bend for a long time after it begins to break; while cast iron, on account of its granular structure, fails suddenly after it begins, and the bending which has previously taken place is not noticed. It is not safe to trust to such general observations for scientific or even practical purposes, but careful observations must be made, so that all the circumstances of the case may be definitely known. It will hereafter be shown that the ultimate resistance to crushing of cast iron is double that of wrought iron, and yet Fairbairn and other English engineers have justly insisted upon the use of wrought iron for tubular and other bridges. For, without considering the comparatively treacherous character of cast iron when heavily loaded, it appears that within the elastic limits (and the structure should not be loaded to exceed that), a wrought iron structure is *stiffer* than a cast iron one of the same dimensions, and will sustain more for a given compression, extension, or deflection.

44. COMPRESSION OF OTHER MATERIALS.—All materials are compressible as well as extensible, and it is generally assumed that their resistance to compression, within the elastic limits, is the same as for extension; but, as has been seen in the previous articles, this is not rigorously correct. Indeed the same piece resists differently under different circumstances, depending upon its temperature, the duration of the strain, and the suddenness with which the force is applied. But these changes are not great, and the mean value of the coefficient of elasticity is sufficiently exact for practical cases.

ULTIMATE STRENGTH.

45. MODULUS FOR CRUSHING.—The modulus of resistance

to crushing is the pressure which is necessary to crush pieces of a material whose length does not exceed from one to five times its diameter, and whose section is unity. The value thus found we call C. It is found by experiment that the resistance of all substances used in the mechanic arts varies very nearly as the section under pressure. Hence, if

P = the crushing force, and

K = the section under pressure, we have

$$P = CK \quad . \quad . \quad . \quad . \quad . \qquad (22)$$

46. MODULUS OF STRAIN.—If the force P is not sufficient to crush the piece, we have for the *strain on a unit of section*

$$C_1 = \frac{P}{K} \quad . \quad . \quad . \quad . \quad .$$

It is necessary to use short pieces in determining the value of C, because long pieces will bend before breaking, and will not be simply crushed, but will break more like a beam.

47. RESISTANCE TO CRUSHING OF CAST-IRON.

TABLE

*Of the results of experiments on the tensile and crushing resistances of cast iron of various kinds, made by Eaton Hodgkinson.**

Description of the iron.	Tensile Strength per square inch. T.	Height of Specimen.	Crushing strength per square inch. C.	Ratio of tenacity to crushing. T : C.	
	Lbs.	Inch.	Lbs.		Mean.
Low Moor Iron, No. 1	12,694	$\frac{3}{4}$ $1\frac{1}{2}$	64,534 56,445	1 : 5·084 1 : 4·446	1 : 4·765
" " No. 2	15,458	$\frac{3}{4}$ $1\frac{1}{2}$	99,525 92,332	1 : 6·438 1 : 5·973	1 : 6·205
Clyde Iron, No. 1....	16,125	$\frac{3}{4}$ $1\frac{1}{2}$	92,869 88,741	1 : 5·759 1 : 5·503	1 : 5·631
" " No. 2....	17,807	$\frac{3}{4}$ $1\frac{1}{2}$	109,992 102,030	1 : 6·177 1 : 5·729	1 : 5·953
" " No. 3....	23,468	$\frac{3}{4}$ $1\frac{1}{2}$	107,197 104,881	1 : 4·568 1 : 4·469	1 : 4·518
Blaenavon Iron, No. 1	13,938	$\frac{3}{4}$ $1\frac{1}{2}$	90,860 80,561	1 : 6·519 1 : 5·780	1 : 6·149
" " No. 2	16,724	$\frac{3}{4}$ $1\frac{1}{2}$	117,605 102,408	1 : 7·032 1 : 6·123	1 : 6·577
" " No. 3	14,291	$\frac{3}{4}$ $1\frac{1}{2}$	68,559 68,532	1 : 4·797 1 : 4·795	1 : 4·796
Calder Iron No. 1....	13,735	$\frac{3}{4}$ $1\frac{1}{2}$	72,193 75,983	1 : 5·256 1 : 5·532	1 : 5·394
Coltness Iron, No. 3..	15,278	$\frac{3}{4}$ $1\frac{1}{2}$	100.180 101,831	1 : 6·557 1 : 6·665	1 : 6·611
Brymbo Iron, No. 1..	14,426	$\frac{3}{4}$ $1\frac{1}{2}$	74,815 75,678	1 : 5·186 1 : 5·264	1 : 5·216
" " No. 3..	15,508	$\frac{3}{4}$ $1\frac{1}{2}$	76,133 76,958	1 : 4·909 1 : 4·963	1 : 4·936
Bowling, No. 2.......	13,511	$\frac{3}{4}$ $1\frac{1}{2}$	76,132 73,984	1 : 5·635 1 : 5·476	1 : 5·555
Ystalyfea, No. 2..... (Anthracite)	14,511	$\frac{3}{4}$ $1\frac{1}{2}$	99,926 95,559	1 : 6·886 1 : 6·585	1 : 6·735
Yniscedwyn, No. 1... (Anthracite)	13,952	$\frac{3}{4}$ $1\frac{1}{2}$	83,509 78,659	1 : 5·985 1 : 5·638	1 : 5·811
" No. 2...	13,348	$\frac{3}{4}$ $1\frac{1}{2}$	77,124 75,369	1 : 5·778 1 : 5·646	1 : 5·712
Stirling, 2d quality..	25,764	$\frac{3}{4}$ $1\frac{1}{2}$	125,333 119,457	1 : 4·865 1 : 4·637	1 : 4·751
" 3d quality..	23,461	$\frac{3}{4}$ $1\frac{1}{2}$	158,653 129,876	1 : 6·762 1 : 5·536	1 : 6·149
Mean.......	16,303	.. ..	88,800 94,730	Mean ratio 1 : 5·64	

* Supplement to Bridges, by Geo. R. Brunell, and Wm. T. Clark. John Weale, London.

In this table the ratio of resistances range from less than 4½ (Clyde, No. 3) to more than 7 (Blaenavon, No. 1). The same experimenter once obtained the ratio of 8.493 from a specimen of Carron iron, No. 2, hot blast;* and the mean of several experiments, made at the same time, gave 6.594. Hence we have, as the mean result of a large number of experiments, that the crushing resistance of cast iron is about 6 times as great as its tenacity; but the extremes are from 4½ to 8½ times its tenacity.

48. RESISTANCE OF WROUGHT IRON TO CRUSHING.—Comparatively few experiments have been made to determine how much wrought iron will sustain at the point of crushing, and those that have been made give as great a range of results as those for cast iron. Wrought iron being fibrous, does not indicate the point of yielding as distinctly as cast iron and other granulous substances.

Hodgkinson gives $C=65000$ †
Rondulet " $C=70800$ ‡
Weisbach " $C=72000$ §
Rankine " $C=30000$ to 40000 ‖

Hence it appears that the crushing resistance of wrought iron is from ½ to ¾ as much as its tenacity.

49. RESISTANCE OF WOOD TO CRUSHING.—The resistance of wood to crushing depends as much upon its state of dryness, and conditions of growth and seasoning, as its tenacity does. The following are a few examples:—

Kind of Wood.	Moderately Dry.	Very Dry.
Ash..................................	8,680	9,360
Oak (*English*)............................	6,480	10,058
Pine (*Pitch*)............................	6,790	6,790

These results, compared with the corresponding numbers in

* Résistance des Matériaux, Morin, p. 95.
† Vose, Handbook of Railroad Construction, p. 127.
‡ Mahan's Civil Engineering, p. 97.
§ Weisbach Mech. and Eng., vol. i., p. 215.
‖ Rankin's Applied Mech., p. 633.

article 22, show that these kinds of wood will resist from $1\frac{1}{2}$ to nearly 2 times as much to tension as to compression.

50. RESISTANCE OF CAST STEEL TO CRUSHING.—Major Wade found the following results from experiments upon the several samples, all of which were cut from the same bar and treated as indicated in the table.*

Specimen.	Length.	Diameter.	Crushing in lbs. per sq. inch.
Not Hardened..................	1·021	0·400	198,944
Hardened, low temper...........	0·995	0·402	354,544
" mean "	1·016	0·403	391,985
" high " for tools for turning hard steel...	1·005	0·405	372,598

51. RESISTANCE OF GLASS TO CRUSHING.—We owe most of our knowledge of the strength of glass to Wm. Fairbairn and T. Tate, Esq. According to their experiments we have the following results for the crushing resistance of specimens of glass whose height varied from one to three times their diameter.

MEAN CRUSHING RESISTANCE OF CUT-GLASS CUBES AND ANNEALED GLASS CYLINDERS.

Description of the Glass.	Weight per Square Inch.	
	Cubes.	Cylinders.
	lbs.	lbs.
Flint Glass........................	13,130	27,582
Green Glass.......................	20,206	31,876
Crown Glass......................	21,867	31,003
Mean....................	18,401	30,153

The ratio of the mean of the resistances is as 1 to 1·6 nearly.

The cylinders were cut from round rods of glass, and hence retained the outer skin, which is harder than the interior, while the cubes were cut from the interior of large specimens. This

* Report on Metals for Cannon, p. 258.

may partially account for the great difference in the two sets of experiments. The cubes gave way more gradually than the cylinders, but both fractured some time before they entirely failed. The cylinders failed very suddenly at last, and were divided into very small fragments. The specimens had rubber bearings at their ends, so as to produce an uniform pressure over the whole section.

52. STRENGTH OF PILLARS.—The strength of pillars for *incipient flexure* has been made the subject of analysis by Euler and others, but practical men do not like to rely upon their results. Mr. Hodgkinson deduced empirical formulas from experiments which were made upon pillars of wood, wrought iron, and cast iron. The experiments were made at the expense of Wm. Fairbairn, and the first report of them was made to the Royal Society, by Mr. Hodgkinson, in 1840. The following are some of his conclusions:—

1st. In all long pillars of the same dimensions, when the force is applied in the direction of the axis, the strength of one which has flat ends is about three times as great as one with rounded ends.

2d. The strength of a pillar with one end rounded and the other flat, is an arithmetical mean between the two given in the preceding case of the same dimensions.

3d. The strength of a pillar having both ends firmly fixed, is the same as one of half the length with both ends rounded.

4th. The strength of a pillar is not increased more than $\frac{1}{7}$th by enlarging it at the middle.

To determine general formulas, bars of the same length and different sections were first used; then others, having constant sections and different lengths; and formulas were deduced from the results. The formulas thus made were compared with the results of experiments on bars whose dimensions differed from the preceding. The following are the results of some of his

EXPERIMENTS ON SQUARE PILLARS.

Length of the bars.	Side of the square.	Crushing weight.	Exponent of the side.
Feet.	Inches.	Lbs.	
10	0·766	1,948	3·57
	1·51	23,025	
10	1·00	4,225	4·17
	1·50	23,025	
7½	1·02	10,236	3·69
	1·53	45,873	
7½	0·50	583	4·08
	1·00	9,873	
5	0·50	1,411	3·67
	1·00	18,038	
2½	0·502	4,216	2·69
	1·00	27,212	
2½	0·502	4,216	3·28
	0·76	15,946	
		Mean....	3·591

The fourth column is computed as follows:—

Suppose that the strengths are as the x power of the diameters, then for the first bar we have

$$\left(\frac{1.51}{0.766}\right)^x = \frac{23025}{1948} \text{ or } 1.987^x = 11.30$$

$$\therefore x = \frac{\log.\ 11.30}{\log.\ 1.987} = 3.57.$$

The others are computed in the same way.

An examination of the table shows that when the square section is the same the strength varies inversely as the length. Thus, of two bars whose cross section is one square inch, the one five feet long is nearly four times as strong as the one ten feet long.

Let l = length of one,
l' = " of other,
d = diameter of first one,
d' = " of the second one, and
x = the power of the length.

Then the strength of the first one is, $P = \text{constant} \times \frac{d^{3.59}}{l^y}$.

" " " second is, $P' = \text{constant} \times \frac{d'^{3.59}}{l'^y}$.

$$\therefore \frac{P}{P'}\left(\frac{d'}{d}\right)^{3.59} = \left(\frac{l'}{l}\right)^{y},$$

in which substitute the values from any two experiments. Thus if we take from the table

$l' = 10$ feet, $d' = 1$ inch, $P' = 4225$ lbs., and
$l = 5$ feet, $d = 1$ inch, and $P = 18038$ lbs., we have

$$\frac{18038}{4225} = 2^{y}$$

$$\therefore y = \frac{\log. 4.2694}{\log. 2} = 2.094.$$

Proceed in a similar way with each of the others and take the mean of the results for the power to be used. In this way was formed the following

TABLE

For the absolute strength of columns.

In which P = crushing weights in gross tons,
d = the external diameter, or side of the column in inches,
d_1 = the internal diameter of the hollow in inches, and
l = the length in feet.

Kind of Column.	Both ends rounded, the length of the column exceeding fifteen times its diameter.	Both ends flat, the length of the column exceeding thirty times its diameter.
	TONS.	TONS.
Solid Cylindrical Columns of cast iron................	$P = 14.9 \frac{d^{3.76}}{l^{1.7}}$	$P = 44.16 \frac{d^{3.55}}{l^{1.7}}$
Hollow Cylindrical Columns of cast iron.....	$P = 13 \frac{d^{3.76} - d_1^{3.76}}{l^{1.7}}$	$P = 44.34 \frac{d^{3.55} - d_1^{3.55}}{l^{1.7}}$
Solid Cylindrical Columns of wrought iron...........	$P = 42 \frac{d^{3.76}}{l^2}$	$P = 133.75 \frac{d^{3.55}}{l^2}$.
Solid Square Pillar of Dantzic oak.................	....	$P = 10.95 \frac{d^4}{l^2}$
Solid square Pillar of red dry deal................	....	$P = 7.81 \frac{d^4}{l^2}$

The above formulas apply only in cases where the length is so great that the column breaks by bending and not by simple crushing. If the column be shorter than that given in the table, and more than four or five times its diameter, the strength is found by the following formula:

$$W = \frac{P.\,CK}{P + \frac{3}{4}CK} \quad - \; - \; - \; - \; - \quad (23.)$$

in which P = the value given in the preceding table,
K = the transverse section of the column in square inches,
C = the modulus for crushing in tons (gross) per square inch, and
W = the strength of the column in tons (gross).*

Experiments have been made upon steel pillars which gave good results.†

53. WEIGHT OF PILLARS.—From the first formula of the preceding table we find

$$d = \frac{P^{\frac{1}{3.76}}\, l^{\frac{0.85}{1.88}}}{14.9^{\frac{1}{3.76}}}.$$

The area of the cross section is $\frac{1}{4}\,\pi\, d^2$, and the volume in inches $= \frac{12}{4}\,\pi\, d^2\, l$.

Cast iron weighs 450 pounds to the cubic foot, hence the

$$\text{weight} = \frac{450}{1728} \times 3 \times \pi d^2 \times l = \frac{450}{576} \times 3.1416 \times \frac{P^{\frac{1}{1.88}}\, l^{\frac{1.79}{0.94}}}{14.9^{\frac{1}{1.88}}},$$

which by reduction gives

$$\text{weight} = 0.0152803\left(P.\, l^{3.58}\right)^{\frac{1}{1.88}} \quad - \; - \; - \; - \quad (24.)$$

If P is given in pounds, this coefficient must be divided by $2240^{\frac{1}{1.88}}$.

If the pillar is hollow the section of the iron is $\frac{1}{4}\,\pi\,(d^2 - d_1^2)$, and if n is the ratio of the diameters, so that $d_1 = n\,d$ this becomes

$\frac{1}{4}\,\pi\, d^2(1 - n^2)$; and its volume in inches $= \frac{12}{4}\,\pi\, d^2\,(1 - n^2)\, l$;

---

* James B. Francis, C.E., has published a set of tables which gives the strength of cast-iron columns, of given dimensions, by means of equation (23), and also by those in the above table.

† *London Builder*, No. 1211.

and its weight in pounds $= \dfrac{450}{1728} \times 3 \times \pi d^2 (1 - n^2)\, l.$

If the value of d from the second equation of the first column in the preceding table, be substituted in the preceding equation, we find the

weight in pounds =

$$\frac{25}{32} \frac{\pi}{\left(2240 \times 13\right)^{\frac{1}{1.88}}} \frac{1 - n^2}{\left(1 - n^{3.76}\right)^{\frac{1}{1.88}}} \left(P.\, l^{1.7}\right)^{\frac{1}{1.88}} \quad (25.)$$

Proceeding in this way with each of the cases given above and we form the following:

TABLE

Of the weights in pounds of pillars in terms of their lengths in feet, and crushing forces in pounds.

Kind of Pillar.	Weight in pounds.	
	Both ends rounded. $l > 15\, d$.	Both ends flat. $l > 30$ d.
Solid Cylindrical Column of cast iron.	$0{\cdot}0101645\, (P.\, l^{3.58})^{\frac{1}{1.88}}$	$0{\cdot}00376452\, (P.\, l^{3.475})^{\frac{1}{1.775}}$
Hollow Cylindrical Columns of cast iron, $d_1 = nd$.	$0{\cdot}0103623 \dfrac{1-n^2}{(1-n^{3.76})^{\frac{1}{1.88}}} \times (P.\, l^{3.58})^{\frac{1}{1.88}}$	$0{\cdot}00375582 \dfrac{1-n^2}{(1-n^{3.55})^{\frac{1}{1.775}}} \times (P.\, l^{3.475})^{\frac{1}{1.775}}$
if $n = 0{\cdot}98$	$0{\cdot}001649014\, (P.\, l^{3.58})^{\frac{1}{1.88}}$	$0{\cdot}000669628\, (P.\, l^{3.475})^{\frac{1}{1.775}}$
if $n = 0{\cdot}95$	$0{\cdot}002549688\, (P.\, l^{3.58})^{\frac{1}{1.88}}$	$0{\cdot}001005503\, (P.\, l^{3.475})^{\frac{1}{1.775}}$
if $n = 0{\cdot}925$	$0{\cdot}003086622\, (P.\, l^{3.58})^{\frac{1}{1.88}}$	$0{\cdot}00120664\, (P.\, l^{3.475})^{\frac{1}{1.775}}$
if $n = 0{\cdot}90$	$0{\cdot}003566870\, (P.\, l^{3.58})^{\frac{1}{1.88}}$	$0{\cdot}00137552\, (P.\, l^{3.475})^{\frac{1}{1.775}}$
if $n = 0{\cdot}875$	$0{\cdot}003982043\, (P.\, l^{3.58})^{\frac{1}{1.88}}$	$0{\cdot}00152392\, (P.\, l^{3.475})^{\frac{1}{1.775}}$
if $n = 0{\cdot}85$	$0{\cdot}004359267\, (P.\, l^{3.58})^{\frac{1}{1.88}}$	$0{\cdot}00165855\, (P.\, l^{3.475})^{\frac{1}{1.775}}$
if $n = 0{\cdot}80$	$0{\cdot}005039546\, (P.\, l^{3.58})^{\frac{1}{1.88}}$	$0{\cdot}00189914\, (P.\, l^{3.475})^{\frac{1}{1.775}}$
$n = 0{\cdot}75$	$0{\cdot}005649247\, (P.\, l^{3.58})^{\frac{1}{1.88}}$	$0{\cdot}00211346\, (P.\, l^{3.475})^{\frac{1}{1.775}}$
Solid Cylindrical Columns of Wrought Iron,	$0{\cdot}00599672\, (P.\, l^{3.58})^{\frac{1}{1.88}}$	$0{\cdot}00201664\, (P.\, l^{3.775})^{\frac{1}{1.775}}$
Square Column of Dantzic Oak.	(Cubic foot weighs 47.24 pounds.)	$0{\cdot}000547291\, P^{\frac{1}{2}}\, l^2$

If the thickness of the metal (t) and the external diameter are given, n may be found as follows: $d-2t$=internal diameter, hence $n=\frac{d-2t}{d}=1-\frac{2t}{d}$. For instance, if the external diameter is 6 inches, and the thickness $\frac{3}{8}$ of an inch, the internal diameter is $5\frac{1}{4}$ inches and $n=\frac{5\frac{1}{4}}{6}=0.875$.

The iron used in the preceding experiments was Low Moor No. 2, whose strength in columns is about the mean of a great variety of English cast iron, the range being about 15 per cent. above and below the values given above.

54. CONDITION OF THE CASTING.—Slight inequalities in the thickness of the castings for pillars does not materially affect the strength, for, as was found by Mr. Hodgkinson, thin castings are much harder than thicker ones, and resist a greater crushing force. In one experiment the shell of a hollow column resisted about 60 *per cent.* more per square inch than a solid one.* But the excess or deficiency of thickness should not in any case exceed 25 per cent. of the average thickness.† Thus, if the average thickness is one inch, the thickest part should not exceed $1\frac{1}{4}$ inch, and the thinnest part should not be less than $\frac{3}{4}$ of an inch.

It is also found that in large castings the crushing strength of the part near the surface does not much exceed that of the internal parts.

55. EXPERIMENTS MADE BY THE NEW YORK CENTRAL RAILROAD COMPANY.—The immediate object of these experiments was to determine the relative values of different sorts and forms of wrought iron of lengths greatly exceeding their diameters, when subjected to longitudinal compression. The pieces were not in all cases broken, nor even materially altered in form by the compressions to which they were subjected, the experiments being generally discontinued as soon as the progressive rate of flexure due to a regularly increased load was ascertained.

The testing machine used in the experiments was designed by C. Hilton, and was made at the Company's carpenter shop, at Albany, by order of the Chief Engineer; its arrangement and all its principal details were afterwards found to be exactly similar to

* Phil Trans., 1857, p. 890.

† Stoney on Strains, vol. ii., p. 206.

those of the machine used for the same purpose by Mr. Eaton Hodgkinson, at Manchester, England (for a description and drawing of which see "Tredgold on the Strength of Wrought and Cast Iron," Weale, London, 1847), with this difference, that the machine made at Albany was of wood, while that used by Mr. Hodgkinson was of iron.

Experiment No. 1.

Made upon a bar of English Crown Iron from the works of Hawkes, Crawshay & Co., Gateshead, planed at both ends and perfectly straight, exactly 8 *feet in length, and of cross section as sketched in Fig.* 14.

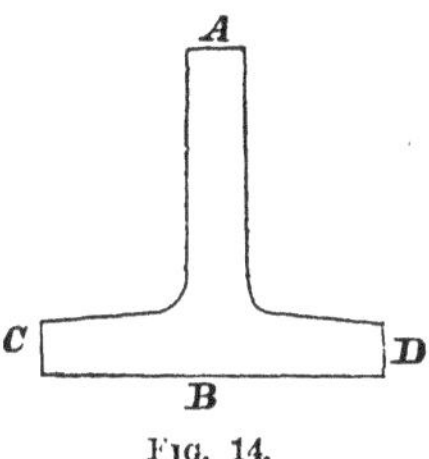

Fig. 14.

Weight applied in lbs.	Deflection in parts of an inch.	Remarks.
2,568	.0	
5,468	.0	
8,468	.033	In direction C D.
10,448	.0625	" "
11,738	.0833	" "
12,778	.15625	" "
13,778	.1875	" "
15,208	.1875	" "
16,758	.25	" "
20,928	.3125	

On the removal of the above recorded load, the bar immediately resumed its original form, having taken no appreciable set.

Being once more placed in the machine the results were :—

Lbs.	Deflections in A B.	Deflections in C D.
20,248	.100	.400
25,968	.300	.600
30,348	Bar bent double in C D.	

EXPERIMENT No. 2.

Made upon a bar of English Crown Iron from the same works, planed at both ends and perfectly straight, exactly 8 *feet in length, and of the cross section shown in Fig.* 15.

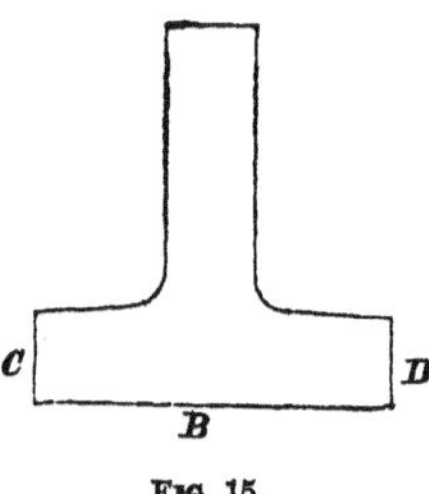

FIG. 15.

Weight applied in lbs.	Deflection in parts of an inch.	Remarks.
2,568	.0	
5,468	.0	
8,178	.05	In direction C D.
10,918	.075	
13,668	.1	
16,468	.125	
19,218	.156	
21,968	.1875	
24,678	.218	
27,478	.25	
30,248	.3125	

No more than 30,248 lbs. was placed upon this bar, the bar being required for use, and the quality of the iron being very soft and easily bent.

No appreciable set was found upon the removal of the above load. On being placed a second time in the machine, and subjected to a load of 19,218 lbs., the flexure was observed to be .125 inch, this weight being left on for 42 hours; on removal a permanent set was observed of .01 inch.

EXPERIMENT No. 3.

Made upon a bar of English Crown Iron from the same works, 8 *feet long, planed at the ends, and cross section sketched below in Fig* 16.

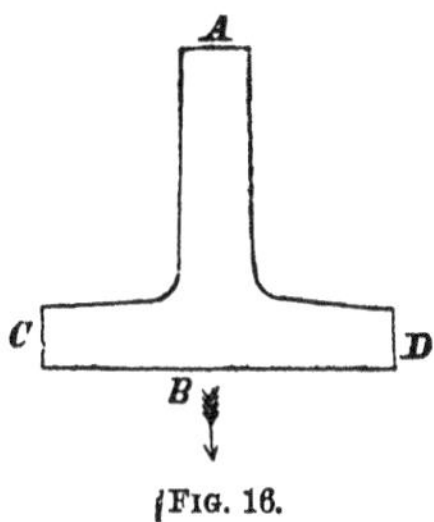

FIG. 16.

Weights applied in lbs.	Deflection in parts of an inch.	Remarks.
2,568	.0	Flexure not appreciable.
5,468	.144	In direction C D.
8,178	.168	" "
10,888	.192	" "
13,638	.242	" "
16,438	.288	" "
19,188	.314	" "
21,938	.360	" "
24,648	.408	" "
27,448	.480	" "
30,218	.612	" "

The bar was found, on removing the weights, to be perfectly straight. When replaced in the machine the results were as follows:—

Weights.	Inches in C D.	Inches in A B.
16,510	.5625	.1875
24,338	.625	.21875
27,638	.6875	.25
30,418	.75	.3125
33,188	.875	.3750
35,948	1.000	.3750

With 35,948 lbs. the bar bent double after four minutes.

EXPERIMENT No. 4.

Made upon a piece of English Crown Iron from the same works, of the section shown in Fig. 17, planed at both ends and exactly 5 feet in length.

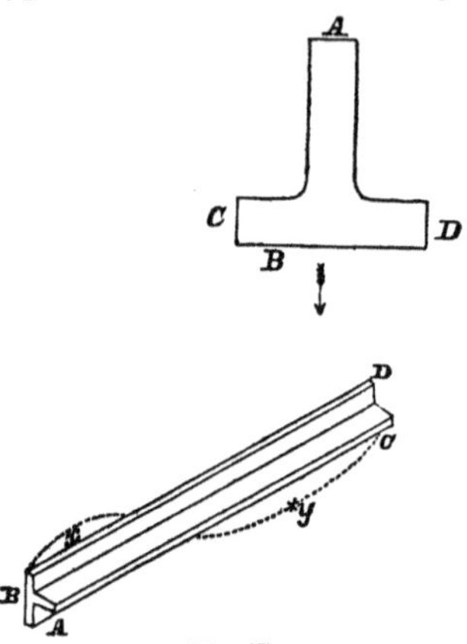

Fig. 17.

Weight in lbs.	Deflection in A B, in parts of an inch.	Deflection in C D, in parts of an inch.
2,568	.00	.0
3,038	.015	.0
4,038	.025	.020
5,468	.060	.050
8,228	.083	.083
10,968	.1	.1
13,678	.1	.125
16,448	.1	.142
19,248	.1	.150
21,998	.083	.166
24,708	.083	.166

On the application of 27,508 lbs. the bar assumed a new form, as shown in the figure; the deflection in the new direction is designated as taking place in x.

Weight in lbs.	Deflection in A B, in parts of an inch.	Deflection in C D, in parts of an inch.	Deflection in x, in parts of an inch.
27,508	0.083	.166	.0875
30,288	.083	.166	.1
33,058	.083	.166	.125
35,818	.083	.225	.1875
39,228	Bar broke at y.		

EXPERIMENT No. 5.

A Bar of Angle Iron, 5 *feet in length, planed at both ends and quite straight, of the cross section shown in Fig.* 18, *furnished by the Albany Iron Works, Troy, of ordinary quality.*

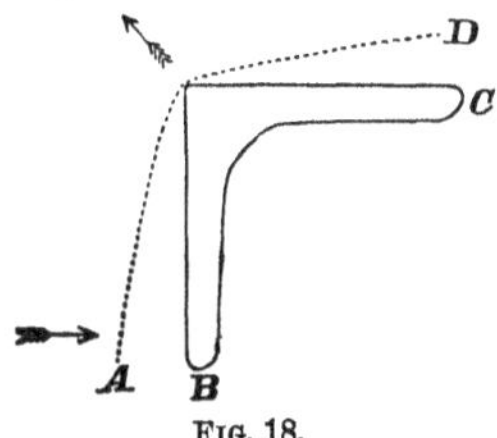

FIG. 18.

Weight in lbs.	Deflection A B.	Deflection C D.
2,568	.04	.05
3.038	.06	.08
4,038	.08	.15
5,468	.375	.25
8,228	.500	.375
10,968	.5625	.375
13,678	.5625	.375
16,448	.625.	.375
19,248	Broke or bent double.	

The general deflection in the direction of the arrow could not be observed until 8,228 lbs. were applied, when it was successively—

.156, .25, .33, .5.

No. 6.

Made upon the bar used in Experiment No. 2. It being presumed that the previous experiment had somewhat weakened this bar, it was determined to break it. The following weights were used:—

Weight in lbs.	Deflection in C D.
22,048	.126
30,398	.5625
35,928	Bar bent nearly double.

56. COMPRESSION OF TUBES.—BUCKLING.—Wrought iron tubes when subjected to longitudinal compressive stresses may yield by crushing like a block, or by bending like a beam, or by buckling. The first takes place when the tube is very short; the second, when it is long compared with the diameter of the tube; and the last, for some length which it is difficult to assign, intermediate between the others.

The appearance of a tube after it has yielded to buckling is shown in Figs. 19 and 20.

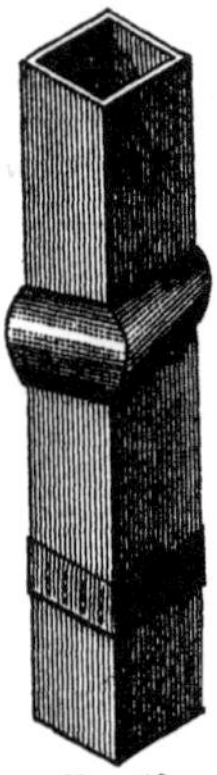

The experiments heretofore made do not indicate a specific law of resistance to buckling; but the following general facts appear to be established:—*

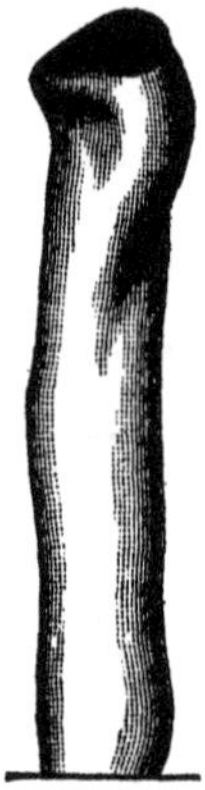

1. The resistance to buckling is always less than that to crushing; and is nearly independent of the length.
2. Cylindrical tubes are strongest; and next in order are square tubes, and then rectangular ones.
3. Rectangular tubes, ▭, are not as strong as tubes of this form ◫. The tubes in bridges and ships are generally rectangular or square.

Fig. 19. Fig. 20.

COLLAPSE OF TUBES.

57. THE RUPTURE OF TUBES which are subjected to great external normal pressure is called "a collapse." The flues of a steam-boiler are subjected to such an external pressure, and in view of the extensive use of steam power, the subject is very important. The true laws of resistance to collapsing were unknown until the subject was investigated by Wm. Fairbairn. Experiments were carefully made, and the results discussed by him with that scientific ability for which he is so noted. They were published in the Transactions of the Royal Society, 1858, and republished in his "Useful Information for Engineers," second series, page 1.

* Civ. Eng. and Arch. Jour., vol. xxviii., p. 28.

The tubes were closed at each end and placed in a strong cylindrical vessel made for the purpose, into which water was forced by a hydraulic press, thus enabling him to cause any desirable pressure upon the outside of the tube. In order to place the tube as nearly as possible in the condition of a flue in a steam-engine, a pipe which communicated with the external air was inserted into one end of the tube. This pipe permitted the air to escape from the tube during collapse.

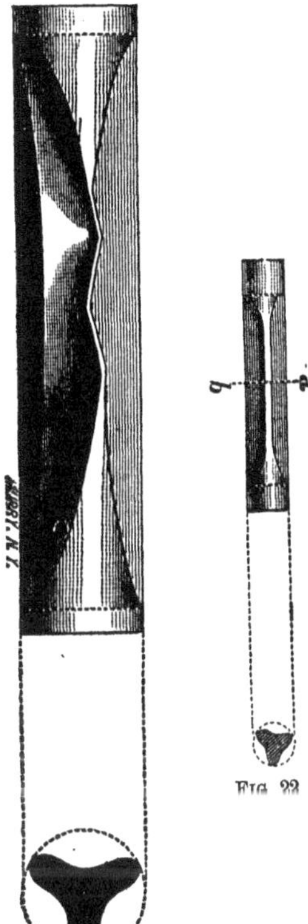

Fig. 22.

Fig. 21.

The vessel, pipe, tube, and their connections were made practically water-tight, and the pressure indicated by gauges.

Fig. 21 shows the appearance and cross-section at the middle of the short tubes after the collapse; and Fig. 22 of a long one. Although no two tubes appeared exactly alike after the collapse, yet the examples which I have selected are good types of the appearances of thirty tubes used in the experiments.

The tubes in all cases collapsed suddenly, causing a loud report. In the first and second tubes the ends were supported by a rigid rod, so as to prevent their approaching each other when the sides were compressed.

The following tables give the results of the experiments :—

TABLE I.

Mark.	No.	Thickness of Plate, inches. t.	Diameter in inches. d.	Length in inches. L.	Pressure of Collapse, lbs. pr. sq. in. of Surface. P.	Product of Pressure and Length. P. L.	Product of the Pressure, Length, and Diameter. P. L. $d.=p$.
A	1	0·043	4	19	170	3230	
B	2	"	"	19	137	2603	10412
C	3	"	"	40	65	2600	10400
D	4	"	"	38	65	2470	9880
E	5	"	"	60	43	2580	10320
F	6	"	"	60	140*	2800	
					Mean	2714	10253
G	7	"	6	30	48†	1440	
H	8	"	"	29	47†	1263	
J	9	"	"	59	32	1888	11328
K	10	"	"	30	52	1560	9360
L	11	"	"	30	65	1950	11700
M	12	"	"	30	85‡	?	
					Mean	1620	10796
N	13	"	8	30	39	1170	9360
O	14	"	"	39	32	1248	9984
P	15	"	"	40	31	1240	9920
					Mean	1219	9754
Q	16	"	10	50	19	950	9500
R	17	"	"	30	33	990	9900
					Mean	970	9700
S	18	"	12·2	58½	11·0	643·7	7850
T	19	"	12	60	12·5	750	9000
V	20	"	"	30	22	662	7920
					Mean	6852	8256

* This tube had two solid rings soldered to it, 20 inches apart, thus practically reducing it to three tubes, as shown in Fig. 23.

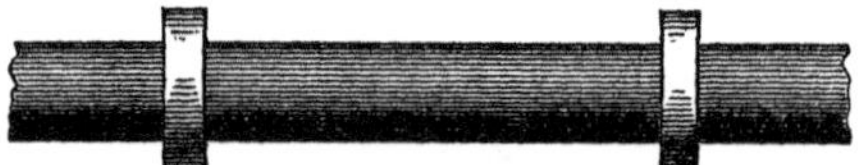

FIG. 23.

† The ends of both were fractured, causing collapse, perhaps before the outer shell had attained its maximum.

‡ A tin ring had been left in by mistake, thus causing increased resistance to collapsing.

58. DISCUSSION OF RESULTS.—By comparing the tubes of the same diameter and thickness, but of different lengths, we see that the long tubes resist less than the short ones; hence, the strength is an *inverse function* of the length, and an examination of the seventh column shows that it is nearly a simple inverse function of the length. The first of the 4-inch tubes is so much stronger than the others, it may be neglected in determining the *law of resistance*, although it differs from a mean of all the others by less than $\frac{1}{5}$ of the mean. An examination of the several cases indicates that we may *safely* assume that *the resistance to collapsing varies inversely as the lengths of the tubes.**

The mean of the results for the several diameters in the last column shows that the resistance diminishes somewhat more rapidly than the diameter increases; but this includes the error, if any, of the preceding hypothesis. As the power of the diameter is but little more than unity, it seems safer to conclude, for all tubes less than 12 inches in diameter, as Fairbairn does, that *the resistance of tubes to collapsing varies inversely as their diameters.*

59. LAW OF THICKNESS.—Experiments were also made to determine the law of resistance in respect to the thickness. Comparatively few experiments were made of this character, but these few gave remarkably uniform results. One of the

---

* A more exact law may be found as follows:—Let P = the compressing force per square inch; C = a constant for any particular diameter and thickness, l = the length, and n the unknown power. Then

$$\mathrm{P} = \frac{C}{l^n} \text{ for one case.}$$

$$\mathrm{P}_1 = \frac{C}{l_1^{\,n}} \text{ for another.}$$

$$\therefore n = \frac{\log. \frac{\mathrm{P}}{\mathrm{P}_1}}{\log. \frac{l_1}{l}}.$$

By means of this equation, and any two experiments in which the thickness and diameter are the same, n may be found, and by using several experiments a series of values may be found from which the most probable result can be obtained. But in this case the *mean* result is so near unity, there is no practical advantage secured by finding it.

tubes (No. 24), was made with a butt joint, as shown in Fig. 24, and the others with lapp joints, as in Fig. 25.

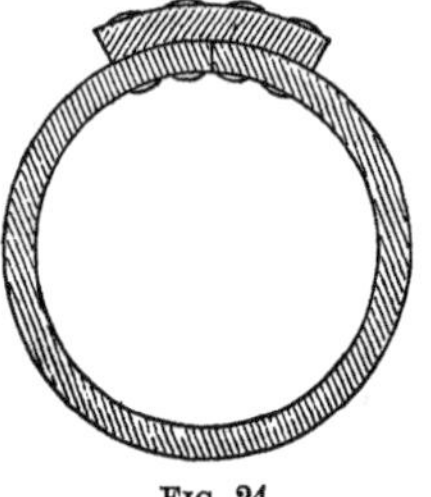

Fig. 24.

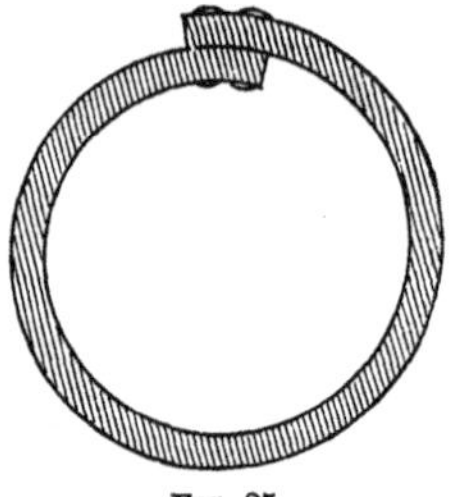

Fig. 25.

The following are the results of the experiments:—

TABLE II.

Mark.	No.	Thickness. t.	Diameter. d.	Length in inches. L.	Pressure per square inch. P.	P. L.	Product. L. P. d. $= p$.
W	21	0·25	9	37	(450)	Uncollapsed.	
X	22	0·25	$18\frac{3}{4}$	61	420	25620	480375
Y	23	0·14	9	37	262	9694	89046
Z	24	0·14	9	37	378	13986	125874
JJ	33	0·125	$14\frac{1}{2}$	60	125	7500	108750

Tubes Nos. 23 and 24 were exactly alike in every respect except their joints; and it appears that the butt joint, No. 24, is 1·41 times as strong as the lap joint, a gain of 41 per cent. But this is a larger gain than is indicated in other cases; for instance, No. 33, which is also a lap joint, offers a greater resistance as indicated in the last column, than No. 23, although the former is not as thick as the latter. Still it seems evident that butt joints are stronger than lap joints, for with the former the tubes can be made circular, and there is no cross strain on the rivets, conditions which are not realized in the latter.

The resistance of the 23d is so small compared with others, it is rejected in the analysis.

We observe that the resistance varies as some power of the thickness; if then C and n be two constants to be determined by experiment, and we use the notation given above, we shall have for the pressure of collapse of one tube.

$$P = \frac{Ct^n}{d\,L} \quad \therefore dPL = p = Ct^n \quad - \ - \ - \ - \ - \quad (24)$$

and for another tube

$$P_1 = \frac{Ct_1^n}{d_1 L_1} \quad \therefore d_1\, P_1\, L_1 = p_1 = Ct_1^n \quad - \ - \ - \quad (25)$$

Hence we have

$$\frac{p}{p_1} = \left(\frac{t}{t_1}\right)^n$$

$$\text{or, } n = \frac{\log.\ p - \log.\ p_1}{\log.\ t - \log.\ t_1} \quad - \ - \ - \ - \ - \ - \ - \ - \quad (26)$$

$$\text{and } C = \frac{p}{t^n} = \frac{p_1}{t_1^n} \quad - \ - \ - \ - \ - \ - \ - \ - \ - \ - \ - \quad (27)$$

TO FIND THE CONSTANTS n AND C.

The mean of the mean of the values of p from Table I. is $p = \frac{1}{5}\,[10253 + 10796 + 9754 + 9700 + 8256] = 9752$ and $t = 0{\cdot}043$.

Using these values and others taken from the preceding tables, and the following values may be found for n:—

In equation (26) make $p = 480375$, $t = 0{\cdot}25$, $p_1 = 9752$, $t_1 = 0{\cdot}043$; and we get

$$n = \frac{\log.\ 480375 - \log.\ 9752}{\log.\ 0{\cdot}25 - \log.\ 0{\cdot}043} = 2{\cdot}200.$$

Similarly, taking $p = 480375$, $t = 0{\cdot}25$, $p_1 = 10253$, $t_1 = 0{\cdot}043$; and we get

$$n = \frac{\log.\ 480375 - \log.\ 10253}{\log.\ 0{\cdot}25 - \log.\ 0{\cdot}043} = 2{\cdot}185.$$

The mean value of p for all but the 12-inch tubes in Table I. is

$$p = \tfrac{1}{4}\,(10253 + 10796 + 9754 + 9700) = 10125;$$

hence, using $p = 125874$, $t = 0{\cdot}14$, $p_1 = 10125$, $t_1 = 0{\cdot}043$; and we get

$$n = \frac{\log.\ 125874 - \log.\ 10125}{\log.\ 0{\cdot}14 - \log.\ 0{\cdot}043} = 2{\cdot}134;$$

and taking $p = 108750$, $t = 0{\cdot}125$, $p_1 = 10125$ and $t_1 = 0{\cdot}043$; we get

$$n = \frac{\log.\ 108750 - \log.\ 10125}{\log.\ 0{\cdot}125 - \log.\ 0{\cdot}043} = 2{\cdot}203,$$

and the mean of these results is, $n = 2{\cdot}18$.

Fairbairn made it 2·19 by including some data which I have rejected as paradoxical; I have also given more weight to those cases which gave nearly uniform results. The difference, however, of 0·01 is too small to seriously affect practical results.

To determine the constant, C, substitute the proper values taken from the preceding tables in equation (27), and we have for four cases the following:—

$$C = \frac{9752}{0{\cdot}043^{2{\cdot}18}} = 9{,}298{,}900.$$

$$C = \frac{480375}{0{\cdot}25^{2{\cdot}18}} = 9{,}864{,}300.$$

$$C = \frac{125874}{0{\cdot}14^{2{\cdot}18}} = 9{,}144{,}000.$$

$$C = \frac{108750}{0{\cdot}125^{2{\cdot}18}} = 10{,}109{,}400.$$

The mean of which is $C = 9{,}604{,}150$. Calling $C = 9{,}600{,}000$ and equation (26) becomes:

$$\mathrm{P} = 9{,}600{,}000\,\frac{t^{2{\cdot}18}}{d.\,\mathrm{L}} \quad - \ - \ - \ - \ - \ - \ - \ - \ - \ - \quad (28)$$

If L be given in feet, so that $\mathrm{L} = 12\mathrm{L}_f$, we have

$$\mathrm{P} = 800{,}000\,\frac{t^{2{\cdot}18}}{d.\,\mathrm{L}_f} \quad - \ - \ - \ - \ - \ - \ - \ - \ - \ - \quad (29)$$

The coefficient, 9,600,000, applies only to the kind of iron used; but the exponent, 2·18, is supposed to be constant for all kinds of iron.

60. FORMULA FOR THICKNESS TO RESIST COLLAPSING.—Equation (28) readily gives the following expression *for finding the thickness in inches of a tube to resist collapsing*:—

$$\log.\,100\,t = \frac{\log.\,\mathrm{P} + \log.\,(d.\,\mathrm{L})}{2{\cdot}18} - 1{\cdot}203 \quad - \ - \quad (30)$$

61. ELLIPTICAL TUBES.—Experiments made upon *elliptical tubes* showed that the preceding formula would give the strength, if the diameter of the circle of curvature at the extremity of the minor axis is substituted for d. The diameter of curvature is $\frac{2a^2}{b}$, in which a is the major and b the minor axis.

Experiments made upon tubes in which the ends were not connected by internal rods, showed that the resistance was inversely as their length.

62. VERY LONG TUBES.—Some experiments were made upon a tube 35 feet long and one 25 feet long. Sufficient pressure was applied to distort them, but not to collapse them, and it was found that equation (28) erred by at least 20 per cent., giving too small an amount. It was, however, very evident that the length was still a very important element in the strength.

63. COMPARISON OF STRENGTH FROM EXTERNAL AND INTERNAL PRESSURE.—Let p be the internal pressure per square inch at which the tube is ruptured, then for tubes of the same thickness and diameter we have from equations (18) and (29), by calling T = 30,000 lbs.,

$$\frac{p}{P} = \frac{1}{13{\cdot}33}\,\frac{L}{t^{1{\cdot}18}}.$$

If $p = P$, then $L = 13{\cdot}33\ t^{1{\cdot}18}$.

If $t = 0{\cdot}25$, then we find $L = 3{\cdot}56$ feet, that is, a tube whose thickness is $\frac{1}{4}$ of an inch, and whose length is 3·56 feet, is equally strong whether subjected to internal or external pressure.

If the tube is so thick that the unequal stretching of the fibres must be considered, then equation (20) must be compared with equation (29), in which case we have :—

$$\frac{p}{P} = \frac{T}{800{,}000} \times \frac{d.\ L_f}{(r+t)t^{1{\cdot}18}}. \qquad - \ - \ - \ - \ - \ - \ (31)$$

If $p = P$, $T = 40{,}000$ lbs., and $2r = d = 4$ inches;
then $2t^{1{\cdot}18} + t^{2{\cdot}18} = \frac{1}{6}\ L_f$.

If $t = \frac{1}{2}$ inch, $L = 5{\cdot}504$ feet.

If $t = 1$ " $L = 15$ feet.

64. RESISTANCE OF GLASS GLOBES TO COLLAPSING.—Fairbairn also determined that glass globes and cylinders followed the same *general law* of resistance. For globes of flint glass he found :

$$P_1 = 28{,}300{,}000\,\frac{t^{1{\cdot}4}}{d^{3{\cdot}4}} \qquad - \ - \ - \ - \ - \ - \ - \ - \ - \ (32)$$

and for cylinders of flint glass :

$$P_1 = 740{,}000\,\frac{t^{1{\cdot}4}}{d\,L} \qquad - \ - \ - \ - \ - \ - \ - \ - \ - \ (33)$$

provided that their length is not less than twice, nor more than six times their diameter. Dividing equation (33) by (28) gives $\frac{P_1}{P} = \frac{0 \cdot 0770}{t^{0 \cdot 78}}$.

If $t = 0 \cdot 043$ in., $\frac{P_1}{P} = 0 \cdot 896$; or the glass cylinder is nearly $\frac{9}{10}$ as strong as the iron one. If they are equally strong, $P = P_1$ $\therefore t = 0 \cdot 0373$ of an inch.

CHAPTER III.

THEORIES OF FLEXURE AND RUPTURE FROM TRANSVERSE STRESS.

65. REMARK.—The ancients seem to have been entirely ignorant of the laws which govern the strength of beams. They made some rude experiments to determine the absolute strength of some solids, especially of stone. They may have recognized some general facts in regard to the strength of beams, such as that a beam is stronger with its broad side vertical than with its narrow side vertical, but we find no trace of any law which was recognized by them. This department of science belongs wholly to modern times. A very brief sketch of the history of its development is given below.*

66. GALILEO'S THEORY.—Galileo was the first writer, of whom we have any knowledge, who endeavored to establish the mathematical laws which govern the strength of beams.† He assumed—

1st. That none of the fibres were elongated or compressed.

2d. When a beam is fixed at one end, and loaded at the other, it breaks by turning about its lower edge, B, Fig. 26; or if it be supported at its ends and loaded at the middle of the length, it would turn about the upper edge; hence every fibre resists tension.

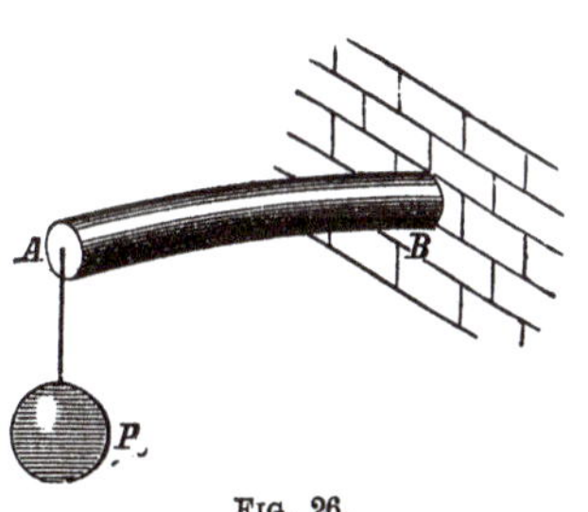

FIG. 26.

3. Every fibre acts with equal energy. From these he readily deduced,—that, when one end is firmly fixed in a wall or other

---

* For a more complete history, see introduction to "*Résistance des Corps Solides*," par Navier. 3d edition. Paris, 1864.

† Opere di Galileo. Bologne, 1856.

immovable mass, the total *resistance of the section* is equal to the sum of all the fibres, or the transverse section, multiplied by the resistance of a unit of section, multiplied by the distance of the centre of gravity from the lower edge. Hence, in a rectangular beam, if

$T =$ the tenacity of the material,
$b =$ the breadth, and
$d =$ the depth of the beam;

the moment of resistance is

$$Tbd \times \tfrac{1}{2}d = \tfrac{1}{2}Tbd^2 \quad - - - - - - - - - \quad (34)$$

67. ROBERT HOOKE'S THEORY.—Robert Hooke was one of the first, and probably the first, to recognize the compressibility of *solids* when under pressure. In 1678 he announced his famous principle, *Ut tensio sic vis;* which he gave in an anagram in 1676, and stated as the basis of the theory of elasticity that the *extensions* or *contractions* were proportional to the forces which produce them, and also that when a bar was bent the material was compressed on the concave side and extended on the convex side.

68. MARRIOTTE'S AND LEIBNITZ'S THEORY.—Marriotte, in 1680, investigated the subject, and finally stated the following principles:—

1st. The material is extended on the convex side and compressed on the concave side.

2d. In *solid rectangular* sections the line of invariable fibres (or *neutral axis*) is at half the depth of the section.

3d. The elongations or compressions increase as their distance from the neutral axis.

4th. The resistance is the same whether the neutral axis is at the middle of the depth or at any other point.

5th. The lever arm of the resistance is $\frac{2}{3}$ of the depth.

We here find some of the essential principles of the resistance to flexure, as recognized at the present day; but the two last are erroneous. As hereafter shown, the neutral axis is at half the depth, and the lever arm is $\frac{2}{3}$ of $\frac{1}{2}$ the depth.

Leibnitz's theory, given in 1684, was the same as Marriotte's.

69. JAMES BERNOUILLI'S THEORY was essentially the same

as Marriotte's, except that he stated that extensions and compressions were not proportional to the stresses. "For," said he, "if it is true, a bar might be compressed to nothing with a finite force." On this point see Article 16. He was the first to give a correct expression for the equation of the elastic curve.

70. PARENT'S THEORY.—Parent, a French academician of great merit, but of comparatively little renown, published, in 1713, as the result of his labors, the following principles, in addition to those of his predecessors:—

1st. The total *resistance* of the compressed fibres equals the total *resistance* of the extended fibres.

2d. The *origin of the moments* of resistance should be on the neutral axis.

By the former of these principles the position of the neutral axis may be found, when the straining force is normal to the axis of the beam; and by the latter he corrected the error of Marriotte and Leibnitz; showing that the ratio of the *absolute* to the *relative* strength is as *six* times the length to the depth, instead of three, as will be shown hereafter.

71. COULOMB, IN 1773, PUBLISHED the most scientific work on the subject of the stability of structures which had appeared up to his time. He deduced his principles from the fundamental equations of statics, and generalized the first of the principles of Parent, which is given above, by saying that *the algebraic sum of all the forces must be zero on the three rectangular axes.* This establishes the position of the neutral axis when the applied forces are oblique to it, as well as when they are normal. He also remarked, that if the proportionality of the compressions and extensions do not remain to the last, or to the point of rupture, the final neutral axis will not be at the centre of the section.

72. MODULUS OF ELASTICITY.—In 1807 Thomas Young introduced the term *modulus of elasticity*, which we have defined as the coefficient of elasticity in Article 5. After this several writers, among them Duhamel, Navier in his early writings, and Barlow in his first work, stated the erroneous principle, *that the sum of the* MOMENTS *of the resistances to compression equalled those for tension.*

73. IN 1824 NAVIER PUBLISHED the lectures which he had given to *l'École des Ponts et Chaussées*, in which he established more clearly those principles of elastic resistance, and resistance to rupture, which have since his day been accepted by nearly all writers. He was the first to show that when the stress is perpendicular to the axis of the beam, *the neutral axis passes through the centre of gravity of the transverse sections.* His most important modifications in the *analysis* was in making $ds = dx$, or otherwise, considering that *for small deflections the tangent of the angle which the neutral axis makes with the original axis of the beam is so small compared with unity that it may be neglected;* and also, that the lever arm of the force remains constant during flexure. These principles we have used in Chapter V. He resolved many problems not before attempted, and became an eminent author in this department of science.

74. THE COMMON THEORY.—The theories of flexure and of rupture which result from these numerous investigations, I will call, for convenience, the *common theory.* It consists of the following hypotheses:—

1st. The fibres on the convex side are extended, and on the concave side are compressed, and there are no strains but compression and extension.

2d. Between the extended and compressed fibres (or elements) there is a surface which is neither extended nor compressed, but retains its original length, and which is called the neutral surface, or in reference to a plane of fibres it is called the *neutral axis.*

3d. The strains are proportional to their distance from the neutral axis.

4th. The transverse sections which were normal to the neutral axis of the beam before flexure, remain normal to the neutral axis during flexure.

5th. A beam will rupture either by compression or extension when the *modulus of rupture* is reached.

6th. The *modulus of rupture* is the strain at the instant of rupture upon a unit of the section which is most remote from the neutral axis on the side which first ruptures. This is called R.

The remainder of this article properly belongs to Chapter

VI., but it is given here so that the reasons for Barlow's theory may be understood.

If a beam ruptures on the convex side, it appears that it ought to break when its tenacity (T) is reached; but it is found by experiment that in this case R always exceeds T. Similarly, it would seem that if it failed by crushing on the concave side, as in the case of rectangular cast-iron beams, R ought to equal C, but experiment shows that in this case R exceeds C; and generally the value of R is always between those of T and C for the same material, being greater than the smaller, and less than the larger. The values of R in the tables were deduced from experiments upon rectangular beams, as will hereafter be shown; and hence, if the common theory is correct, R should equal the value of the lesser resistance, whether it be for compression or extension; but it does not. This discrepancy between theory and the results of experiment * has led Barlow to investigate the subject further, and it has resulted in a new theory which he calls "Resistance to Flexure"—an expression which I consider unfortunate, as it does not express his idea. "Longitudinal Shearing" would express his idea better, as will appear from the following article:—

75. BARLOW'S THEORY.—According to the common theory

* Mosley's Mech. and Arch., p. 557. "The elasticity of the material has been supposed to be perfect up to the instant of rupture, but the extreme fibres are strained much beyond their elastic limits before rupture takes place, while the fibres near the neutral axis are but slightly strained, and hence the law of proportionality is not maintained, and the position of the neutral axis is changed, and the sum of the moments is not accurately $\frac{RI}{d_1}$ (see equation 171). To determine the influence of these modifications we must fall back upon experiment, and it has been found *in the case of rectangular beams* that the error will be corrected if we take $\frac{1}{m}$ T (= R) instead of T, where m is a constant depending upon the material."

Weisbach, vol. ii., 4th ed., p. 68, foot-note says, "Excepting as exhibiting approximately the laws of the phenomena, the theory of the strength of materials has many practical defects."

In the Report of the Ordnance Department, by Maj. Wade, p. 1, it says:— "A trial was made with cylindrical bars in place of square ones. These generally broke at a point distant from that pressed, and the results were so anomalous that the use of them was soon abandoned. The formula by which the strength of round bars is computed appears to be not quite correct, for the unit of strength in the round bars is uniformly much higher than in the square bars cast from the same iron."

the resistance at a section is the same as if the fibres acted independently of each other, and the transverse section remained normal to the neutral axis. But Barlow correctly considers that in order to keep the transverse sections normal to the neutral axis, the consecutive planes of fibres must *slide* over each other, and to this movement they offer a resistance.

He presented his view to the Royal Society (Eng.), in 1855, and it has since been published in the *Civil Engineer and Architect's Journal*, vol. xix., p. 9, and vol. xxi., p. 111.* The subject is there discussed in a very able and thorough manner, and although he may have failed to establish his theory, by not taking into account all the incidents which exist at the instant of rupture, yet the results of his analysis seem to agree more nearly with the results of experiment than those obtained by any other theory heretofore proposed.

It is admitted in this theory that a beam will rupture when the stress upon any fibre equals its tenacity, or its resistance to compression, as the case may be. But, on the other hand, when the adjacent fibres are unequally strained, as they are in the case of flexure, it requires a greater stress to produce this strain than it would if the fibres acted independently, according to the previously assumed law. This, Barlow makes evident from the following example:—

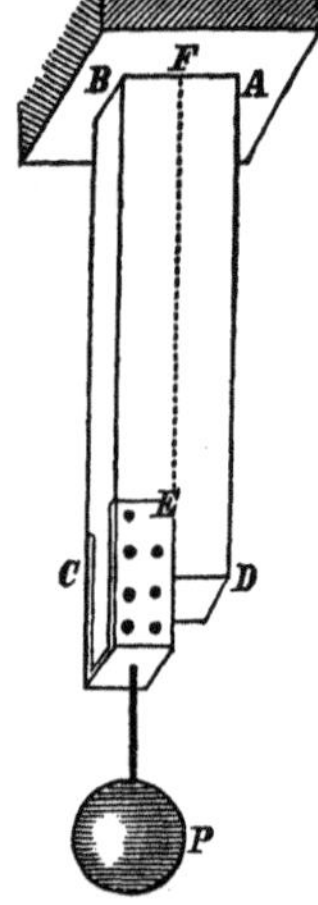

FIG. 27

If a weight, P, Fig. 27, is suspended on a prismatic bar, BCEF, all the fibres will be equally strained, and hence equally elongated.

But if the bar ABCD be substituted for the former, and the weight P acts upon a part of the section, as shown in the figure, it is evident that all the fibres will not be equally strained, and hence will not be equally elongated; and if the force P was just sufficient to rupture the bar FBCE, it will not be sufficient to rupture the bar ABCD, although P acts directly upon the same section, for the cohe-

* *Civ. Eng. and Arch. Jour.*, Vol. xix., p. 9, Barlow says that the strength of a cast-iron rectangular bar, as found from existing theory, cannot be recon-

sion of the particles along FE will not permit the fibres next to that line to be elongated as much as if the part AFED were removed; and these fibres will act upon those adjacent, and so on, till they produce an effect upon BC. From this we see that it takes a greater weight than P acting upon the section EC to produce a strain T per unit of section, when the part ADEF is added. It is also evident that if the section of ABCD is twice as great as FBCE, it will not take twice P to rupture the fibres on the side BC.

A phenomenon similar to this takes place in transverse strain. One side is compressed and the other elongated; and the fibres less strained aid those which are more strained by virtue of the cohesion which exists between them, and it takes a greater force to cause a strain, T, longitudinally upon the fibres than it would if there were no cohesion.

There is, then, at the time of the rupture of a beam, a tensile strain on the extended fibres, and a compressive strain on the other fibres, and a longitudinal shearing strain between the fibres, due to cohesion. These remarks will, I trust, enable the reader to understand the difference between the "Common Theory" and "Barlow's."

Barlow's Theory consists of the following hypotheses:—

1st. The fibres or elements on the convex side are extended, and on the concave side compressed.

2d. There is a neutral surface, as in the common theory.

3d. The tensive and compressive strains on a fibre are proportional to the distance of the fibre from the neutral axis.

4th. That in addition to these there is a "Resistance to flexure" or longitudinal shearing strain, which consists of the following principles:—

a. It is a strain in addition to the direct extensive and compressive forces, and is due to the lateral cohesion of the adja-

---

ciled with the results of experiment if the neutral axis be at the centre of the sections. He then proceeded to show by experiment that the neutral axis is at the centre, and then remarked that the formula commonly used for a beam supported at the ends and loaded in the middle, or $W = \frac{2}{3}\frac{Tbd^2}{l}$ did not give half the actual strength if T is the tenacity of the iron. He then proceeds to point out a new element of strength, which he calls "Resistance to Flexure."

cent surfaces of fibres or particles, and to the elastic reaction which ensues when they are unequally strained.

b. It is evenly distributed over the surface, and consequently within the limits of its operation its centre of action will be at the centre of gravity of the compressed or of the extended section. This force for solid beams Barlow calls ϕ, and for T or I sections, or open-built beams, it is easily deduced from the following principle:—

c. It is proportional to and varies with the *inequality* of strain between the fibres nearest the neutral axis and those most remote.

From this it appears that if d' is the depth of the horizontal flanges of the I section, and d_1 the distance of the most remote fibre from the neutral axis, then the *resistance to flexure of the flanges* will be $\phi \frac{d'}{d_1}$ and similarly for other forms.

5. Sections remain normal to the neutral axis during flexure.

6. Rupture of solid beams takes place when the strain on a unit of section is $T + \phi$, or $C + \phi$, whichever is smaller, or rather, whichever value is first reached.

76. REMARKS UPON THE THEORIES.—For scientific purposes it is desirable to determine the correct theory of the strength of beams, but the phenomena are so complex that it is not probable that a single general theory can be found which will be applicable to all the irregular forms of beams used in practice. Although Barlow's theory appears plausible, yet according to principle *c* the *resistance to flexure*, ϕ, cannot be uniform over the surface, as stated in principle *b*, because the proportionality of the elongations and compressions do not continue up to the point of rupture. The common theory is faulty beyond what has already been said in the I section; for in the upper and lower portions the strains on all the fibres are not proportional to their distances from the neutral axis, to realize which the material should be continuous; and Barlow's theory is defective in the same case, on account of the peculiar strains upon the fibres at the angles where the parts join. For rupture, then, we can use these theories to ascertain general facts, and make the results safe in practice by using a proper coefficient of safety; but for flexure the common theory is sufficiently exact if the elastic limit is not passed, and this is fortunate as the conditions of stability should be founded on the elastic properties rather than on the ultimate strength of the material. For the rupture of rectangular beams the common theory will be sufficiently exact if the value of R is used instead of T or C in the formulas.

the strains should have been carried as near to the point of rup ture as possible.

78. POSITION DETERMINED ANALYTICALLY.—We know from statics that the algebraic sum of all the forces on each of the rectangular axes must be zero for equilibrium; hence, if the deflecting forces are normal to the axis of the beam, *the sum of the resistances to compression must equal those for tension.*

1st. Suppose that the coefficient of elasticity for compression equals that for tension. Then will the compressions and extensions be equal at equal distances from the neutral axis. In Fig. 28, let R_c be the strain on a unit of fibres most remote from the neutral axis on the compressed side, and $d_c =$ the distance of the most remote fibre on the same side; then,

$$\frac{R_c}{d_c} = s = \text{strain at a unit's distance from the neutral axis.}$$

Let k_1, k_2, k_3, &c., be the sections of fibres on one side of the neutral axis, at distances of

y_1, y_2, y_3, &c., from the axis, and

k', k'', k''', &c., and y', y'', y''', &c., corresponding quantities on the other side.

$$\text{Then } s\,(k_1y_1 + k_2y_2 + k_3y_3 + \&c.) = s\,(k'y' + k''y'' + k'''y''' + \&c.),$$
$$\text{or, } k_1y_1 + k_2y_2 + k_3y_3 + \&c. - (k'y' + k''y'' + k'''y''' + \&c.) = 0,$$
$$\text{or, } \Sigma ky = 0 \qquad (35)$$

or the neutral axis passes through the centre of gravity of the sections.*

If the resistance to compression is greater than for tension, the neutral axis will be nearer the compressed side than when they are equal.

2. Suppose that the coefficient of elasticity is not the same for tension as for compression.

---

* The analytical expression for the ordinate to the centre of gravity is

$$\overline{Y} = \frac{k_1y_1 + k_2y_2 + \&c.\ k'y' + k''y'' + \&c.}{k_1 + k_2 + \&c. + k' + k'' + \&c.}, \text{ or } \overline{Y} = \frac{\iint ydydx}{\int ydx}.$$

$$\text{If } \iint ydydx = 0,\ \overline{Y} = 0.$$

Let Fig. 28 represent the beam. Suppose that the sections CM and EF were parallel before deflection. If through N, the point where EF intersects the neutral axis, KH is drawn parallel to CM, the ordinates between EF and KH will represent the elongations on one side, and the compressions on the other, for those fibres whose original length was LN.

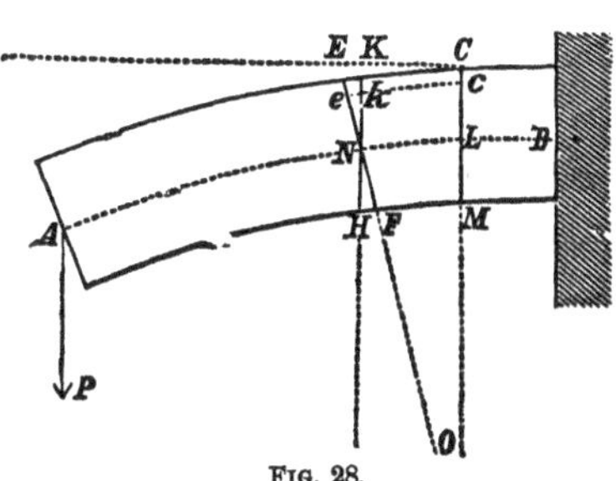

FIG. 28.

Let $l =$ LN,

$\lambda = ke =$ the elongation of a fibre at k;

$p =$ a pulling or pushing force which would produce λ;

$y =$ N$k =$ distance of any fibre from the neutral axis;

$k =$ section of any fibre;

$E_t =$ coefficient of elasticity for tension; and

$E_c =$ " " " compression.

From equation (3) we have,

$$p = \frac{E_t k\lambda}{l} \quad - - - - - - - - - - - - -$$

But λ is directly proportional to its distance from the neutral axis; hence, if c be a constant quantity, whose value may or may not be known, we shall have $\lambda = cy$

$$\therefore p = \frac{cE_t ky}{l} \quad - - - - - - - - \quad (36)$$

Or, if we adopt the same notation as in the preceding case, we shall have for the total force tending to produce extension,

$$\Sigma p = \frac{cE_t}{l}(k_1y_1 + k_2y_2 + k_3y_3 + \&c.) \quad - - \quad (37)$$

Similarly for compression

$$\Sigma p = \frac{cE_c}{l}(k'y' + k''y'' + k'''y''' + \&c.) \quad - \quad (38)$$

Placing these equal to each other and we have,

$$E_t(k_1y_1 + k_2y_2 + k_3y_3 + \&c.) = E_c(k'y' + k''y'' + k'''y''' + \&c.)$$

or, in the language of the integral calculus,

$$E_t \Sigma_0^{y} ydydx = E_c \Sigma_{-y}^{0} ydydx, \quad - - - - - \quad (39)$$

in which y is an ordinate and x an abscissa. Equation (39) enables us to find the position when the form of section is known. In most cases, however, the reduction is not easily made.

Example.—Suppose the sections are rectangular.

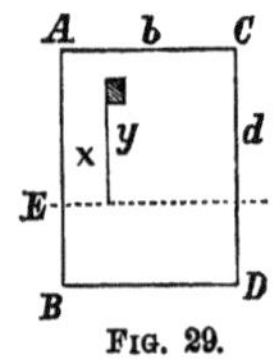

FIG. 29.

Let $b = AC$,
$d = AB$,
$\frac{E_t}{E_c} = a$, and
$y = AE$ for the superior limit.

Then equation (39) becomes

$$a\int_0^b\int_0^y ydydx = \int_0^b\int_0^{d-y} ydydx,$$ which reduced becomes

$$ab\frac{y^2}{2} = \frac{b}{2}\left[d-y\right]^2$$

$$\therefore y = \frac{d}{1+\sqrt{a}} \quad \text{- - - - - - - - - - -} \quad (40)$$

If $a = 1,\ y = \frac{d}{2}$

$a = \infty,\ y = 0$
$a = 0,\ y = d.$

If y is known in equation (40), the ratio of the coefficients may easily be found; for, we have from (40)

$$a = \left(\frac{d-y}{y}\right)^2 = \frac{E_t}{E_c} \quad \text{- - - - - - - -} \quad (41)$$

3d. Suppose that the deflecting force is not perpendicular to the axis, and $E_c = E_t = E$.

Let $\theta =$ the angle which P makes with the axis of the beam Fig. 30;

$P_1 = P \cos\theta =$ the component of P in the direction of the axis of the beam;

$P_2 = P \sin\theta =$ the component of P perpendicular, to the axis of the beam;

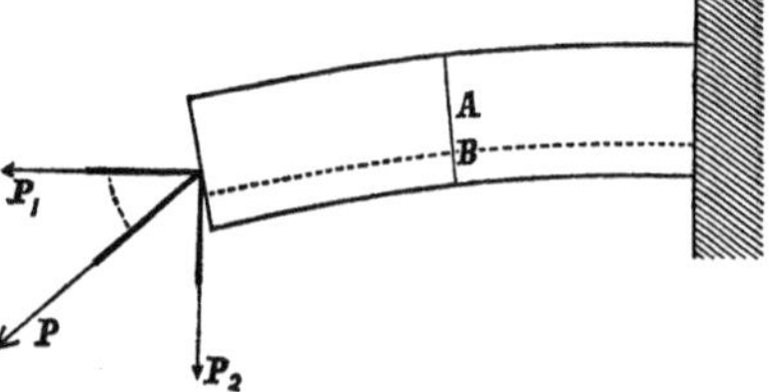

FIG. 30.

$h =$ the distance of the neutral axis from the centre of gravity of the section AB, and K = the transverse section.

The whole force of compression equals the whole force of extension, equations (37) and (38).

$$\therefore \text{P} \cos\theta + \frac{c\text{E}}{l}\iint_0^v y\, dy\, dx = \frac{c\text{E}}{l}\iint_{-v}^0 y\, dy\, dx$$

But the ordinate to the centre of gravity is (see foot-note on page 84),

$$h = \frac{-\iint_0^v y\, dy\, dx + \iint_{-v}^0 y\, dy\, dx.}{\text{K}}$$

$$\therefore \text{ P } \cos\theta = \frac{c\text{E}}{l}\text{K}h$$

$$\text{or } h = \frac{\text{P}l}{c\text{EK}} \cos\theta \quad - \; - \; - \; - \; - \; - \; - \; - \qquad (42)$$

If $\theta = 90°$, $h = 0$ as before found.

If $\theta = 0$ there is no neutral axis, for the force coincides with the axis of the beam. The equation would show the same result, if the value of $c = \frac{\lambda}{y} = \frac{l}{\rho}$, equation (45), were substituted in the formula, for then ρ would be infinite, for $c = 0$, and h becomes infinite.

4th. Let the law of resistance be according to Barlow's *theory of flexure*, and the deflecting forces normal to the axis of the beam.

Using the same notation as before, also

d_1 = the distance of the most remote fibre from the neutral axis, and

ϕ = the coefficient of longitudinal shearing stress.

Then $\varphi \int_0^v y\, dx$ = the resistance to shearing for tension,

and $\varphi \int_{-v}^0 y\, dx$ = the resistance to shearing for compression,

and, proceeding as we did to obtain equation (39), we have

$$\frac{\text{T}}{d_1}\iint_0^v y\, dy\, dx + \varphi \int_0^v y dx = \frac{\text{T}}{d_1}\iint_{-v}^0 y\, dy\, dx + \varphi \int_{-v}^0 y\, dx. \quad (43)$$

Examples.—Let the sections be rectangular, b = the breadth, d = the depth. Then (43) becomes

$$\tfrac{1}{2}\,\text{T}d_1 + \phi\, d_1 = \frac{\text{T}}{2d_1}(d - d_1)^2 + \phi\,(d - d_1)$$

$$\text{or, } \left(\phi + \frac{\text{T}d}{2d_1}\right)(2d_1 - d) = 0\,;$$

$$\therefore d_1 = \tfrac{1}{2}d \cdot \text{ or, } d_1 = -\frac{\text{T}d}{2\phi},$$

the former only of which is admissible.

If the section is a double T, as in Fig. 31, with the notation as in the figure, ϕ will be used in finding the resistance of the vertical rib, and according to Article 75, $\phi \frac{d'}{d-d_1}$ of the lower flange, and $\phi \frac{d_2}{d_1}$ of the upper flange.

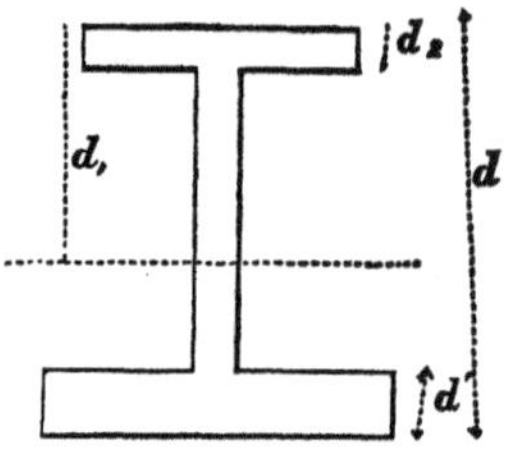

FIG. 31.

It appears from these several cases that the neutral axis passes near the centre of gravity in most practical cases, and it will be assumed that it passes through the centre unless otherwise stated.

CHAPTER IV.

SHEARING STRESS.

79. GENERAL STATEMENT.—Two kinds of shearing stress are recognized—longitudinal and transverse—both of which have been defined in Article 2. Materials under a variety of circumstances are subjected to this stress—such as, rivets in shears; the rivets in riveted plates; pins and bolts in spliced joints; beams subjected to transverse strains; bars which are twisted; and, in short, all pieces which are subjected to any kind of distorsive stress in which all parts are not equally strained. In the first examples above enumerated, all parts of the section are supposed to be equally strained. Shearing may take place in detail, as when plates or bars of iron are cut with a pair of shears, when only a small section is operated upon at a time; or it may be so done as to bring into action the whole section at a time, as in the process of punching holes into metal, where the whole surface of the hole which is made is supposed to resist uniformly.

80. MODULUS OF SHEARING.—The modulus of resistance to shearing is the resistance which the material offers per unit of section to being forced apart when subjected to a shearing stress.

This we call S_s. The resistance for both kinds of shearing has been found to vary directly as the section; so that if K = the area of the section subjected to this stress the total resistance will be

$$K. S_s.$$

The value of S_s has been found for several substances, the principal of which are as follows:—

METALS.

	Ss in lbs. per square inch.
Fine cast steel *	92,400
Rivet steel †	64,000
Wrought iron *	50,000
Wrought-iron plates punched ‡	51,000 to 61,000
Wrought iron hammered scrap punched §	44,000 to 52,000
Cast iron	30,000 to 40,000
Copper ‖	33,000

WOOD.

With the fibres.

White pine	480
Spruce	470
Fir ¶	592
Hemlock **	540
Oak	780
Locust	1,200

Across the fibres.

Red pine	500 to 800
Spruce	600
Larch ††	970 to 1,700
Treenails, English oak ‡‡	3,000 to 5,000

It will be seen from these results that the shearing strength of wrought iron is about the same as its tenacity; of cast steel it is a little less than its tenacity; of cast iron it is double its tenacity, and about $\frac{3}{8}$ its crushing resistance; and of copper it is about $\frac{3}{8}$ its tenacity.

The following table, which gives the results of some experiments upon punching plate iron, illustrates the law of resistance, and gives the value of Ss for that material.

* Weisbach Mech. and Eng., vol. i., p. 407.
† Kirkaldy's Exp. Inq., p. 71.
‡ Proc. Inst. Mech. Eng. England, 1858, p. 76.
§ Proc. Inst. Mech. Eng. England, 1858, p. 73.
‖ Stoney on Strains, vol. ii., p. 284.
¶ Barlow on the Strength of Materials, p. 24.
** Engineering Statics, Gillespie, p. 33.
†† Tredgold's Carpentry, p. 42.
‡‡ Murray on Shipbuilding Wood and Iron, p. 94.

TABLE

*Of Experiments on Punching Plate Iron.**

Diameter of the hole.	Thickness of the plate.	Sectional area cut through.	Total pressure on the punch.	Pressure per square inch of area.
Inch.	Inch.	Square inch.	Tons.	Tons.
0·259	0·437	0·344	8·384	24·4
0·500	0·625	0·982	26·678	27·2
0·750	0·625	1·472	34·768	23·6
0·875	0·875	2·405	55·500	23·1
1·000	1·000	3·142	77·170	24·6

These results give for the value of S_s from 51,000 lbs. to 61,000 lbs. The total resistance varies nearly as the cylindrical surface of the hole.

APPLICATIONS.

81. PROBLEM OF A TIE-BEAM.—*To find the relation between the distance AB, Fig. 32, and the depth of a rectangular beam below the notch, so that the total shearing strength shall equal the total tenacity.*

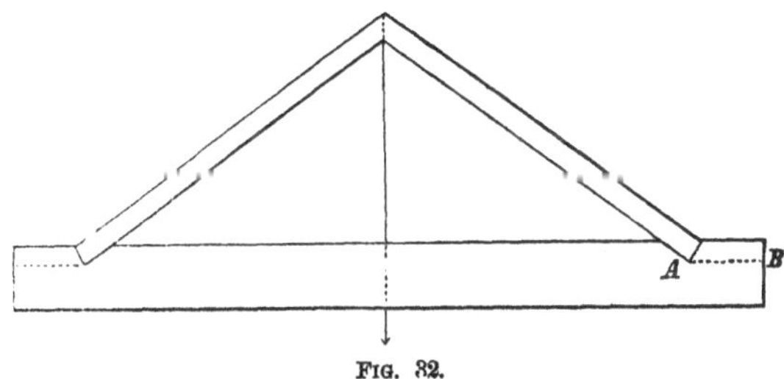

FIG. 32.

Let h = AB = the distance of the bottom of the notch from the end,

d = the remaining depth of the beam,

k = the section of AB,

K = the section below A,

T = the modulus of tenacity, and

S_s = the modulus of shearing strength:

Then the condition requires that

$$\text{TK} = S_s k, \text{ but } k : \text{K} :: h : d$$

$$\therefore \frac{k}{\text{K}} = \frac{h}{d} = \frac{\text{T}}{S_s}$$

$$\therefore h = \frac{\text{T}d}{S_s}$$

* Proceedings Inst. Mech. Eng., 1858, p. 76.

Example.—For Oak $\frac{T}{Ss} = \frac{12000}{780} = 15\frac{1}{3}$ nearly; hence AB should be about $15\frac{1}{3}$ times the remaining depth.

82. RIVETED PLATES.—*Given the diameter of the rivets; it is required to find the distance between them from centre to centre, so that the strength of the rivets for a single row shall equal the strength of the remaining iron in the plates.*

Let d = the diameter of the rivets,
c = the distance between them from centre to centre,
k = the section of the rivet,
K = the remaining section of the plate, and
t = the thickness of the plate.

For iron $T = Ss$; hence, proceeding as above, and we have

$$\frac{k}{K} = \frac{\frac{1}{4}\pi d^2}{t(c-d)} = 1 \therefore c = \frac{0.7854d^2}{t} + d.$$

Examples.—If $t = \frac{1}{4}$ inch, and $d = \frac{1}{2}$ inch;
then $c = 1.2854$, inch,

and $\frac{c-d}{c} = 0.61$.

If $t = \frac{1}{4}$ inch, and $d = \frac{3}{8}$ inch; then $c = 0.8238$ and $\frac{c-d}{c} = 0.544$, which is nearly the value given by Fairbairn for the strength of single riveted plates. See Article 27. To insure this strength the rivet should fit tightly in the hole.

83. LONGITUDINAL SHEARING IN A BENT BEAM.—When a beam is subjected to a transverse stress, we have already seen, Articles 74 and 75, that the fibres are unequally strained, and hence are unequally elongated and compressed. This cannot be done without producing a shearing stress between the adjacent elements or fibres, as shown in Figs. 27 and 28. This shearing strain rarely overcomes the cohesion of the particles, but if they were held only by friction it might overcome that. To illustrate this latter idea, suppose several boards from ordinary lumber are placed upon each other, and the whole supported at the ends in any convenient way. When in this condition draw several straight lines across the pile, perpendicular to the central board. Then deflect the whole by a weight at the middle, or in any other convenient manner, and it will be observed that the lines are no longer straight, but broken, and the general direction does not remain normal to the axis or central board. In the experiment the top layer, instead of being shortened as in a solid beam, retains its length by overcoming the friction between the top board and the one immediately under it. The friction, whether it be much or little, represents the shearing stress in a beam.

The elongations and compressions of the fibres in a bent beam being proportional to their distances from the neutral axis, Article 74, it follows that the shearing stress is evenly distributed over the cross section; and that, beginning at the axis, the *total* shearing stress increases uniformly with the distance from the axis. In a beam which is bent by forces perpendicular to the axis the shearing resistances to compression and tension form a couple whose arm is the distance between the centre of the compressed section and the centre of the extended section. This resistance in bent beams is generally elastic. The coefficient of elasticity for this case for fibrous bodies has not been determined.

84. TRANSVERSE SHEARING IN BENT BEAMS.—Quite analogous to the preceding case is that of transverse shearing in a beam which is bent by external forces. Referring to Fig. 28, in order that the weight P should be sustained by the horizontal beam, there is necessarily a vertical force, or a vertical component of forces in the beam, and it is the same at all sections between A and B. This is easily shown by the principles of mechanics.

In order to simplify the problem, suppose that all the bending forces are in a plane, and let

P, P_1, P_2, &c., be the bending forces,

F, F_1, F_2, &c., be the forces in a beam, each of which is the resultant of all the forces concurring at that point,

$\alpha, \alpha_1, \alpha_2$, &c., the angles which P, P_1, &c., make with the axis of x,

a, a_1, a_2, &c., the angles which F, F_1, F_2, &c., make with the axis of x, and y an axis perpendicular to x.

Then the principles of statics give the following equations:

$$\Sigma P \cos \alpha + \Sigma F \cos a = 0,$$

$$\Sigma P \sin \alpha + \Sigma F \sin a = 0,$$

$$\Sigma(Py \cos \alpha - Px \sin \alpha) + \Sigma\,(Fy \cos a - Fx \sin a) = 0.$$

Let x coincide with the axis of the beam, and let all the forces be vertical; or $\alpha = 90°$ or $180°$; then

$$\left.\begin{array}{llr}(1) & - - - & \Sigma F \cos a = 0 \\ (2) & - - - & \Sigma \pm P + \Sigma F \sin a = 0 \\ (3) & - - - & \Sigma \pm Px + \Sigma Fy \cos a - \Sigma Fx \sin a = 0\end{array}\right\} (44a)$$

The first of these equations shows that the sum of the resisting forces parallel to the axis is zero; or that the total compression equals the total tension. This is equation (35) in another form. The second shows that the sum of the bending forces equals the sum of the vertical components of the resisting forces. If we let Ss represent a strain as well as a modulus, this equation becomes $\Sigma P = \Sigma F \sin a = Ss$, which is the result sought.

This is as far as it is necessary to carry the investigation in this connection; but it may be well to show the use of the third equation. If we use a resultant moment for each of the above sets of moments the equation becomes

$$P'x' - x'' \Sigma F \sin a = F'y',$$

$$\text{or, } P'x' - x'' Ss = F'y'; \text{ but } Ss = \Sigma P = P',$$

$$\therefore P'(x' - x'') = F'y';$$

hence the shearing stress forms a couple with the applied force,—or resultant of applied forces. This equation under the form

$$\Sigma Px = \Sigma Fy$$

is an essential one in Articles 86 and 136.

Examples of transverse shearing stress. The second of equations (44*a*), as reduced is,

$$Ss = \Sigma P.$$

1. Let a beam be uniformly loaded over its whole length, and supported at its ends as in Fig. 42,

and let w = the load on a foot of length,
l = the length of the beam,
$V = \frac{1}{2}wl$ = the amount sustained at each support,
x = any distance from either end; then
wx = the load on the length x; and the expression for the shearing stress becomes

$$Ss = \tfrac{1}{2}wl - wx,$$

which is the equation of a straight line (see Fig. 100). Its value is greatest for $x = 0$, for which it is $\frac{1}{2}wl = \frac{1}{2}W$; and is zero for $x = \frac{1}{2}l$.

2. Suppose the beam is supported at its ends, and has a weight at the middle of its length.

Let P = the weight, and the other notation as before; then $V = \frac{1}{2}P$, and $s = \frac{1}{2}P - 0$ to the middle, and beyond the middle $Ss = \frac{1}{2}P - P = -\frac{1}{2}P$; and hence it is constant over its whole length.

3. If the beam sustains a uniform load, and also a uniformly increasing load from one end, as in Fig. 98, in which W_1 is the total load which increases;

we have $$Ss = V - wx - W_1\frac{x^2}{l^2}$$

$$= \tfrac{1}{2}wl + \tfrac{1}{3}W_1 - wx - W_1\frac{x^2}{l^2}$$

85. SHEARING RESISTANCE TO TORSION.—When a piece is twisted there is a tendency in one section to slip over the adjacent one, and the corresponding resistance constitutes a shearing strain. It is least at the axis, and increases gradually as we proceed from it.

CHAPTER V.

FLEXURE.

82. ELASTIC CURVE.

WHEN a beam is bent by a transverse strain, equilibrium is established between the external and internal forces; or, to be more specific, all the external forces to the right or left of any transverse section are held in equilibrium by the elastic resistances of the material in the section. When in this state the curve assumed by the neutral axis is called the *elastic curve.*

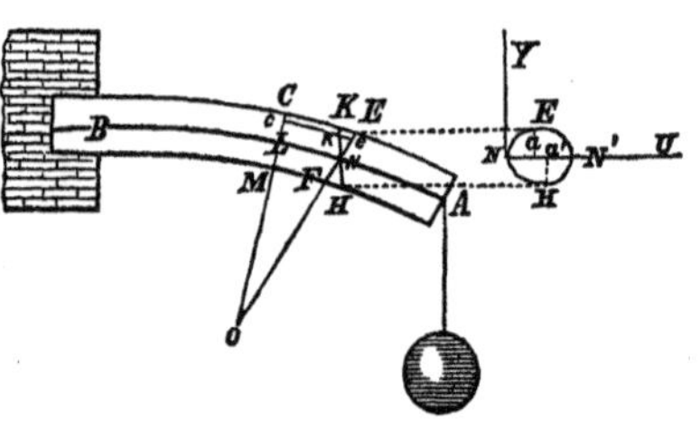

FIG. 33.

To find the general equation of the elastic curve, let Fig. 33 represent a beam, fixed at one end, or supported in any manner, and deflected by a weight, P, or by any number of forces. AB is the neutral axis. Take the origin of coördinates at B (or at any other point on the neutral axis), and let x be horizontal and coincide with the axis of the beam before flexure, y vertical and u perpendicular to the plane of xy. The transverse sections CM and EF being consecutive and parallel before flexure, will meet after flexure, if sufficiently prolonged in some point, as o. Through N draw KH parallel to CM; then will ke be the elongation of a fibre whose original length was ck. We have the following notation:—*

$dx = \mathrm{LN} =$ the distance between consecutive sections,

$y = \mathrm{N}e =$ any ordinate of the surface,

$u = \mathrm{N}a$ or $\mathrm{N}a'$,

$b = \mathrm{NN}' =$ the limiting value of u,

$f(y,u) =$ equation of the transverse section,

* Several of the more important problems of this chapter are solved in Articles 93 to 103, without the use of the calculus.

$dy\, du =$ the transverse section of a fibre,

$b = NN' =$ limiting value of u,

$\rho = ON =$ the radius of curvature at N,

$p =$ the force necessary to elongate any fibre an amount equal to λ when applied in the direction of its length,

$\lambda = ke$,

$I =$ the moment of inertia of the section,

$E =$ the coefficient of elasticity of the material, which is supposed to be the same for extension and compression,

$\Sigma Px =$ a general expression for the moment of applied forces.

We suppose that the strain is within the elastic limit, and establish the algebraic equation on the condition that the sum of the moments of the applied or deflecting forces equals the sum of the moments of the resisting forces. We also assume that the neutral axis coincides with the centre of the transverse sections of the beam.

By the similarity of the triangles LON and kNe, we have

$$ON : Ne :: LN : ke, \text{ or } \rho : y :: dx : \lambda$$

$$\therefore \lambda = \frac{y}{\rho} dx \quad - \; - \; - \; - \; - \; - \; - \quad (45)$$

The force necessary to produce this elongation is (see equation (3)),

$$p = E\, dy\, du \frac{\lambda}{dx};$$

which becomes, by substituting λ from (45),

$$p = \frac{E}{\rho} y\, dy\, du \quad - \; - \; - \; - \; - \; - \; - \; - \quad (46)$$

and the moment of this force is found by multiplying it by y;

$$\therefore py = \frac{E}{\rho} y^2\, dy\, du \quad - \; - \; - \; - \; - \; - \; - \; - \; - \quad (47)$$

The total moment of all the resisting forces to extension and compression is found by integrating (47) so as to include the whole transverse section, and this will equal the sum of the moments of the applied forces:

$$\therefore \frac{E}{\rho}\left[\int_0^b \int_0^{+u} y^2 dy\, du + \int_0^b \int_{-u}^{0} y^2 dy\, du\right] = \Sigma Px$$

$$\text{or } \frac{E}{\rho}\int_0^b\int_{-u}^{+u} y^2 dy\, du = \Sigma Px \quad - - - - - - - - \quad (48)$$

The quantity $E\int\int y^2 dy\, du$, which depends upon the form of the transverse section and nature of the material, is called the *moment of flexure.*

The quantity $\int\int y^2 dy\, du$, when taken between limits so as to include the whole transverse section, is called the moment of inertia of the surface.* Calling this I and equation (48) becomes

$$\frac{EI}{\rho} = \Sigma Px, \quad - - - - - - - - - - - - \quad (49)$$

which is the equation of the *elastic curve.*

An exact solution of equation (49) is not easily obtained in practice, except in a few very simple cases; but when the deflection is small an approximate solution, which is generally comparatively simple and always sufficiently exact, is easily found.

$$\text{We have, } \rho = \frac{(dx^2 + dy^2)^{\frac{3}{2}}}{d^2y\, dx} = ^* \frac{dx^2\left(1 + \frac{dy^2}{dx^2}\right)^{\frac{3}{2}}}{d^2y}$$

$$= \frac{dx^2}{d^2y} \text{ nearly, since for small deflections}$$

$\frac{dy}{dx}$ (which is the tangent of the angle which the tangent line to the curve makes with the axis of x) is small compared with unity, and hence may be omitted. Hence equation (49) becomes

$$EI\frac{d^2y}{dx^2} = \Sigma Px, \quad - - - - - - - - - - - \quad (50)$$

which is the general *approximate* equation of the neutral axis.

87. THE MOMENT OF INERTIA† of all transverse sections of a prismatic beam, is constant, and hence I is constant for prismatic beams.

---

* See Appendix.

† See Appendix.

For a rectangle, as Fig. 34, we have

$$\mathrm{I} = \int_0^b \int_{-\frac{1}{2}d}^{+\frac{1}{2}d} y^2 dy\, du = \tfrac{1}{12} bd^3 \quad - \quad (51)$$

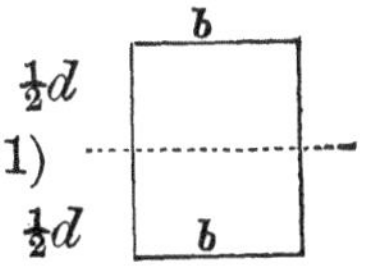

Fig. 34.

For a circle, the origin of coördinates being at the centre;

$$y = r \sin \theta$$
$$dy\, du = rd\,r\,d\,\theta$$

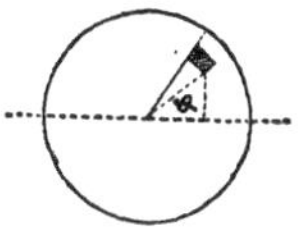

Fig. 35.

$$\therefore \mathrm{I} = \int_0^r \int_0^{2\pi} r^3 dr\, d\theta \sin^2\theta = \tfrac{1}{4}\pi r^4 \quad - \quad - \quad (52)$$

SPECIAL CASES OF PRISMATIC BEAMS.

88. Required the equation of the neutral axis, amount of deflection, and slope of the curve of a prismatic beam, when slightly deflected, and subjected to certain conditions as follows:—

89. Case I.—Suppose a horizontal beam is fixed at one extremity and a weight P rests upon the free extremity; required the equation of the neutral axis and the total deflection.

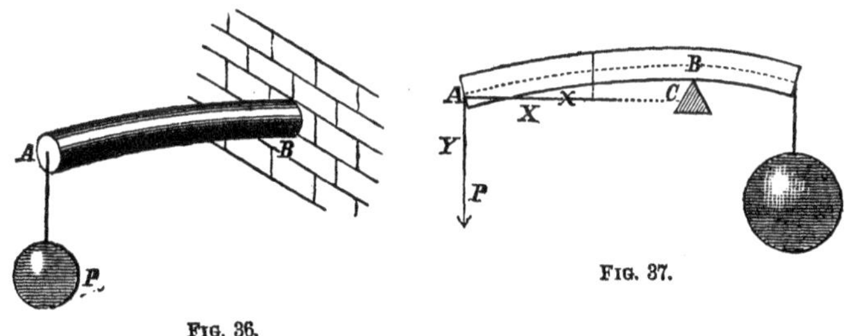

Fig. 36.

Fig. 37.

The beam may be fixed by being imbedded firmly in a wall, as in Fig. 36, or by resting on a fulcrum and having a weight applied on the extended part, which is just sufficient to make the curve horizontal over the support, as in Fig. 37. The latter case more nearly realizes the mathematical condition of fixedness. In either case let

$l = AB =$ the length of the part considered,
$i =$ the inclination of the curve at any point, and
$\Delta = BC =$ the total deflection.

Take the origin of coördinates at the free end, A; x horizontal, y vertical and positive downwards. The moment of P on any section distant x from A is Px, which is the second member of equation (50) in this case. Hence the equation becomes

$$EI\frac{d^2y}{dx^2} = Px \quad - \; - \; - \; - \; - \; - \; - \; - \; - \; - \; - \quad (53)$$

Multiply both members by the dx and integrate, and we have

$$EI\frac{dy}{dx} = \tfrac{1}{2}Px^2 + C_1 \quad - \; - \; - \; - \; - \; - \; - \; - \; - \quad (54)$$

When the deflections are small, the length of the beam remains sensibly constant, hence for the point B, $x = l$; and at the fixed end $\frac{dy}{dx} = 0$. Substitute these values in equation (54), and we find $C_1 = -\frac{1}{2}Pl^2$, and (54) gives

$$\frac{dy}{dx} = \frac{P}{2EI}(x^2 - l^2) = \text{tang}\, i \quad - \; - \; - \; - \; - \; - \quad (55)$$

The integral of equation (55) is

$$y = \frac{P}{6EI}(x^3 - 3\,l^2x) + C_2$$

But the problem gives $y = 0$ for $x = 0 \therefore C_2 = 0$;

$$\therefore y = \frac{P}{6EI}(x^3 - 3\,l^2\,x) \quad - \; - \; - \; - \; - \; - \; - \; - \; - \; - \quad (56)$$

which is the equation of the neutral axis, and may be discussed like any other algebraic curve.

The greatest slope is at A, to find which make $x = 0$ in equation (55)

$$\therefore \text{tang}\, i \text{ (at the free end)} = -\frac{Pl^2}{3EI}$$

The greatest distance between the curve and the axis of x is at B, to find which make $x = l$ in equation (56), and we have

$$y = \Delta = -\frac{Pl^3}{3EI} \quad - \; - \; - \; - \; - \; - \; - \; - \; - \; - \; - \quad (57)$$

If y were positive upward, everything else remaining the same, the second member of equation (53) would have been negative, for it is a principle in the differential calculus that when the curve is concave to the axis of x, the second differential coefficient and the ordinate must have contrary signs. This would make tang i and Δ positive. It will be a good exercise for the student to solve this and other problems by taking the origin of coördinates at different points, only keeping x horizontal and y vertical. For instance, take the origin at B; at C; at the point where the free end of the beam was before deflection; at the middle of the beam; or at any other point.

Example.—If $l = 5$ ft., $b = 3$ in., $d = 8$ in., $E = 1,600,000$ lbs., and $P = 5,000$ lbs.; required the slope at the free end and at the middle, and the maximum deflection.

90. CASE II.—SUPPOSE THAT THE BEAM IS FIXED AT ONE END, IS FREE AT THE OTHER, AND HAS A LOAD UNIFORMLY DISTRIBUTED OVER ITS WHOLE LENGTH.—The beam may be fixed as before, as shown in Figs. 38 and 39.

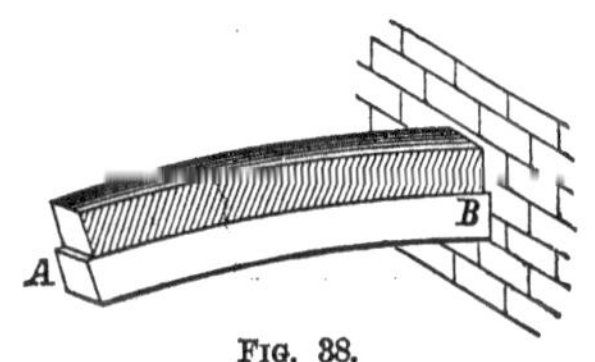

FIG. 38.

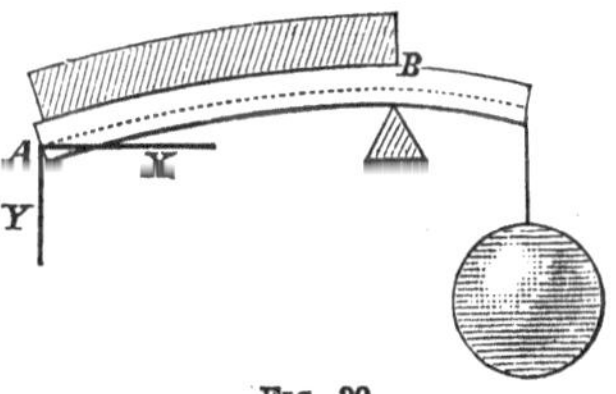

FIG. 39.

Let w = the load on a unit of length. This load may be the weight of the beam, or it may be an additional load.

$W = wl$ = the total load.

Take the origin at A.

Then wx = the load on a distance x, and

$\frac{1}{2} wx^2$ = the moment of this load on a section distant x from A.

Hence equation (50) becomes

$$EI \frac{d^2y}{dx^2} = \frac{1}{2} wx^2 \quad - \; - \; - \; - \; - \; - \; - \; - \; - \; - \quad (58)$$

$$\therefore \frac{dy}{dx} = \frac{w}{6EI}(x^3 - l^3) = \text{tang } i \quad - \; - \; - \; - \; - \quad (59)$$

$$\therefore y = \frac{w}{24EI}(x^4 - 4 l^3 x) \quad - \; - \; - \; - \; - \; - \; - \quad (60)$$

$$\text{and } \Delta = -\frac{wl^4}{8EI} = -\frac{Wl^3}{8EI} \quad - \; - \; - \; - \; - \; - \; - \; - \quad (61)$$

In which $\frac{dy}{dx} = 0$ for $x = l \; \therefore \; C_1 = -\frac{wl^3}{6EI}$,
$y = 0$ for $x = 0 \; \therefore \; C_2 = 0$, and
$y = \Delta$ for $x = l$.

If the origin of coördinates were at the fixed end, ΣPx in the first case would be $P(l - x)$, and in the second $\frac{w}{2}(l - x)^2$. The student may reduce these cases and find the constants of integration. This case may be further modified for practice by taking the origin of coördinates at different points.

91. CASE III.—LET THE BEAM BE FIXED AT ONE END AND A LOAD UNIFORMLY DISTRIBUTED OVER ITS WHOLE LENGTH, AND A WEIGHT ALSO APPLIED AT THE FREE END.—This is a combination of the two preceding cases, and is represented by Figs. 36 and 37, in which the weight of the beam is the uniform load.

$$\therefore EI\frac{d^2y}{dx^2} = Px + \tfrac{1}{2}wx^2;$$

$$\text{and } \Delta = -\frac{l^3}{3EI}(P + \tfrac{3}{8}W) \quad - \; - \; - \; - \; - \; - \; - \; - \quad (62)$$

hence the deflection of a beam fixed at one end and free at the other, and uniformly loaded, is $\frac{3}{8}$ as much as for the same weight applied at the free end.

92. CASE IV.—LET THE BEAM BE SUPPORTED AT ITS ENDS AND A WEIGHT APPLIED AT ANY POINT.—Figs. 40 and 41 represent the case.

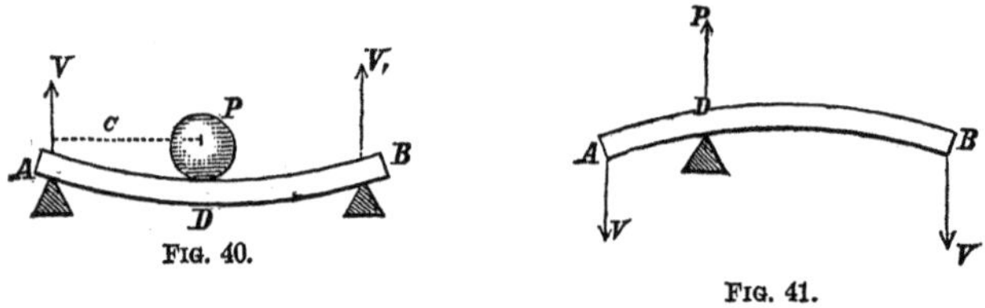

FIG. 40.
FIG. 41.

Let the reaction of the supports be V and V_1. Take the

origin at A over the support, and let AD $= c =$ the abscissa of the point of application of P.

$$\text{Then, } V = \frac{l-c}{l} P, \text{ and } V_1 = \frac{c}{l} P.$$

The case is the same as if a beam rested on a support at D, and weights equal to V and V_1 were suspended at the ends.

For the part AD, equation (50) becomes :—

$$EI \frac{d^2y}{dx^2} = -Vx = -\frac{l-c}{l} Px; \quad - \ - \ - \ - \quad (63)$$

$$\therefore \frac{dy}{dx} = -\frac{P(l-c)}{2lEI} x^2 + C_1; \quad - \ - \ - \ - \ - \ - \ - \quad (64)$$

$$\text{and } y = -\frac{P(l-c)}{6lEI} x^3 + C_1 x + (C_2 = 0); \quad - \ - \ - \quad (65)$$

in the last of which, $y = 0$ for $x = 0$ $\therefore$ $C_2 = 0$ as indicated.

For the part DB, the origin of coördinates remaining at A, we have :—

$$EI \frac{d^2y}{dx^2} = -Vx + P(x-c) = Pc \frac{x-l}{l} = -V_1(l-x); \quad (66)$$

$$\therefore \frac{dy}{dx} = \frac{Pc}{2lEI} (x^2 - 2lx) + C'; \quad - \ - \ - \ - \ - \ - \ - \ - \quad (67)$$

$$\text{and } y = \frac{Pc}{12lEI} (2x^3 - 6lx^2) + C'x + C''. \quad - \ - \ - \ - \quad (68)$$

To find the constants, make $x = c$ in equations (64) and (67) and place them equal to each other; do the same with (65) and (68); and also observe that in (68) $y = 0$ for $x = l$. These conditions establish the three following equations :—

$$-\frac{Pc^2(l-c)}{2lEI} + C_1 = \frac{Pc^2}{2lEI} (c - 2l) + C'$$

$$-\frac{Pc^3(l-c)}{6lEI} + C_1 c = \frac{Pc^3}{12lEI} (2c - 6l) + C'c + C''$$

$$0 = -\frac{Pcl^2}{3EI} + C'l + C''$$

From these we find

$$C_1 = \frac{Pc}{6EIl}(c^2 + 2l^2 - 3cl)$$

$$C' = \frac{Pc}{6EIl}(+ c^2 + 2l^2)$$

$$C'' = -\frac{Pc^3}{6EI}$$

Hence, for the part AD we have

$$EI\frac{d^2y}{dx^2} = -\frac{l-c}{l}Px,$$

$$\frac{dy}{dx} = -\frac{P(l-c)}{2lEI}x^2 + \frac{Pc}{6EIl}(c^2 + 2l^2 - 3cl)$$

$$\text{or, } \frac{dy}{dx} = \frac{P}{6EIl}\left[(-3l + 3c)x^2 + c^3 + 2cl^2 - 3c^2l\right] \quad - \quad - \quad - \quad (69)$$

$$y = \frac{P}{6EIl}\left[(c-l)x^3 + (c^2 + 2l^2 - 3cl)cx\right] \quad - \quad - \quad - \quad (70)$$

To find the maximum deflection, if c is greater than $\frac{1}{2}l$, make $\frac{dy}{dx} = 0$ in (69) and find x; then substitute the value thus found in equation (70). If $c < \frac{1}{2}l$ make $\frac{dy}{dx} = 0$ in equation (67) and substitute the value thus found in equation (68).

If D is at the middle of the length, make $c = \frac{1}{2}l$ in equations (63), (69), and (70); and we have for the curve AD

$$EI\frac{d^2y}{dx^2} = -\frac{1}{2}Px, \quad - \quad - \quad - \quad - \quad - \quad - \quad - \quad - \quad - \quad - \quad (71)$$

$$\frac{dy}{dx} = \frac{P}{16EI}(l^2 - 4x^2),$$

$$y = \frac{P}{48EI}(3l^2x - 4x^3), \quad - \quad - \quad - \quad - \quad - \quad - \quad - \quad - \quad - \quad (72)$$

$$\text{and } \Delta = -\frac{Pl^3}{48EI} \text{ (if } x = \tfrac{1}{2}l \text{ in (72))} \quad - \quad - \quad - \quad - \quad (73)$$

The greatest stress is at the centre, and the maximum moment is found by making $x = \frac{1}{2}l$ in the second member of equation (71). Hence, *the maximum moment is*

$$\tfrac{1}{4}Pl \quad - \quad - \quad - \quad - \quad - \quad - \quad - \quad - \quad - \quad - \quad - \quad - \quad - \quad - \quad (73a)$$

In this case the curve DB is of the same form as AD, but its equation will not be of the same form unless the origin of co-ordinates be taken at the other extremity of the beam.

93. CASE V.—SUPPOSE THAT A BEAM IS SUPPORTED AT OR NEAR ITS EXTREMITIES, AND THAT A LOAD IS UNIFORMLY DISTRIBUTED OVER ITS WHOLE LENGTH.

No account is made of the small portion of the beam (if any) which projects beyond the supports. The distance between the supports is the length of the beam which is considered.

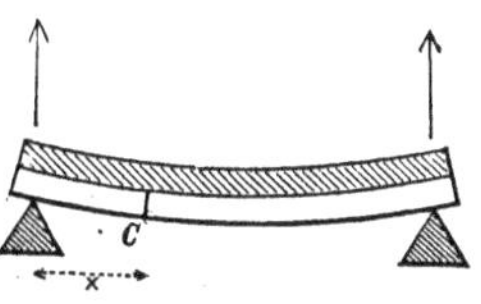

FIG. 42.

Let the notation be the same as in the preceding cases.

Then $V = \frac{1}{2}wl = \frac{1}{2}W =$ the weight sustained by each support;

$Vx = \frac{1}{2}wlx =$ the moment of V on any section, as c;

wx is the load on x, and the lever arm of this load is the distance from its centre to the section c, or $\frac{1}{2}x$; hence its moment is $\frac{1}{2}wx^2$, and the total moment is the difference of the two moments. Hence equation (50) becomes

$$EI\frac{d^2y}{dx^2} = \frac{1}{2}w(-lx + x^2); \quad - \; - \; - \quad (74)$$

$$\therefore \frac{dy}{dx} = \frac{w}{24EI}(-6lx^2 + 4x^3 + l^3);$$

$$y = \frac{w}{24EI}(-2lx^3 + x^4 + l^3x); \; - \; - \; - \quad (75)$$

and if $x = \frac{1}{2}l$ in (75), $y = \Delta = \frac{5}{384}\frac{wl^4}{EI} = \frac{5}{384}\frac{Wl^3}{EI}$ - - (76)

In these equations $\frac{dy}{dx} = 0$ for $x = \frac{1}{2}l, \therefore C_1 = \frac{wl^3}{24EI}$;

and $y = 0$ for $x = 0, \therefore C_2 = 0$.

94. CASE VI.—LET THE BEAM BE SUPPORTED AT ITS ENDS, UNIFORMLY LOADED, AND ALSO A LOAD MIDWAY BETWEEN THE SUPPORTS.

This case is a combination of the two preceding ones, and

may be represented by Fig. 40; for the weight of the beam may be the uniform load. Hence,

$$EI\frac{d^2y}{dx^2} = -\tfrac{1}{2}Px + \tfrac{1}{2}wx^2 - \tfrac{1}{2}wlx \quad - \; - \; - \; - \quad (77)$$

$$\triangle = \frac{l^3}{48EI}\left[P + \frac{5}{8}W\right] \quad - \; - \; - \; - \; - \; - \; - \; - \quad (78)$$

Experiments on the deflection of beams are generally made in accordance with this case. If the beam be rectangular, we have from equation (51),

$$I = \frac{1}{12}bd^3, \text{ which in (78) gives}$$

$$\triangle = \frac{l^3}{4Ebd^3}\left[P + \frac{5}{8}W\right] \quad - \; - \; - \; - \; - \; - \; - \quad (79)$$

$$\therefore E = \frac{l^3}{4\triangle bd^3}\left[P + \frac{5}{8}W\right] \quad - \; - \; - \; - \; - \; - \quad (80)$$

In making an experiment to determine E, the beam is weighed, and that portion of it which is between the supports and unbalanced will be W, and all the quantities except E may be directly measured. If E be known, we may measure or assume all but one of the remaining quantities, and solve the equation to find the remaining quantity, as the following examples will illustrate:—

95. *Examples.*—1. If a rectangular beam, 5 feet long, 3 inches wide, and 3 inches deep, is deflected $\frac{1}{10}$ of an inch by a weight of 3,000 lbs. applied at the middle; required the coefficient of elasticity. E = 20,000,000.

2. If $b = 2$ inches, $d = 4$ inches, and $l = 6$ feet, the weight of the beam 144 lbs., and a weight P=10,000 lbs. placed at the middle of the beam deflects it $\frac{1}{4}$ an inch; required E. E = 14,580,000 lbs.

3. A joist, whose length is 16 feet, breadth 2 inches, depth 12 inches, and coefficient of elasticity 1,600,000 lbs., is deflected $\frac{1}{2}$ inch by a weight in the middle; required the weight; the weight of the beam being neglected.
Ans. P = 1,562 lbs.

4. An iron rectangular beam, whose length is 12 feet, breadth $1\frac{1}{2}$ inches, coefficient of elasticity 24,000,000 lbs., has a weight of 10,000 lbs. suspended at the middle; required its depth that the deflection may be $\frac{1}{480}$ of its length.
Ans. 8.8 in.

5. A rectangular wooden beam, 6 inches wide and 30 feet long, is supported at its ends. The coefficient of elasticity is 1,800,000 lbs.; the weight of a

cubic foot of the beam is 50 lbs.; required the depth that it may deflect 1 inch from its own weight.

How deep must it be to deflect $\frac{1}{480}$ of its length?

6. A cylindrical beam, whose diameter is 2 inches, length 5 feet, weight of a cubic inch of the material 0.25 lb., is deflected $\frac{3}{8}$ of an inch by a weight $P =$ 3,000 lbs. suspended at the middle of the beam. Required the coefficient of elasticity.

To solve this substitute $I = \frac{1}{4}\pi r^4$ (equation (52)) in equation (80). This gives

$$E = \frac{l^3}{12\,\Delta\,\pi r^4}\left[P + \frac{5}{8}W\right]$$

7. Required the depth of a rectangular beam which is supported at its ends, and so loaded at the middle that the elongation of the lowest fibre shall equal $\frac{1}{1400}$ of its original length. (Good iron may safely be elongated this amount.)

Equations (49) and (73a) become $\frac{EI}{\rho} = \frac{1}{4}Pl \mathrel{.\,\dot{.}\,.} \rho = \frac{4EI}{Pl}$. In this substitute the value of I, equation (51), and it becomes

$$\rho = \frac{Ebd^3}{3Pl}. \quad \text{By the problem find } \rho = 700d$$

$$\therefore d = \sqrt{\frac{2100Pl}{Eb}}.$$

8. Required the radius of curvature at the middle point of a wooden beam, when $P = 3,000$ lbs.; $l = 10$ ft.; $b = 4$ in.; $d = 8$ in.; and $E = 1,000,000$ lbs.

Equations (49) and (73a) give $\rho = \frac{EI}{\frac{1}{4}Pl} = \frac{1,000,000 \times \frac{1}{12} \times 4 \times 8^3}{\frac{1}{4} \times 3,000 \times 10 \times 12} = 1,896$ inches.

9. Let the beam be iron, supported at its ends. Let $b = 1$ in., $d = 2$ in., $l =$ 8 ft., $E = 20,000,000$ lbs. Required the radius of curvature at the middle when the deflection is $\frac{1}{4}$ of an inch. Use eqs. (49) and (73) for P at the middle.

$$\therefore \rho = \frac{EI}{\frac{1}{4}Pl} = \frac{E.I}{\frac{1}{4} \times \frac{48E.I.\Delta.l}{l^3}} = \frac{l^2}{12\Delta} = 3,840 \text{ inches};$$

from which it appears that it is independent of the breadth and depth.

10. The centrifugal force caused by a load moving over a deflected beam may be found from the expression $\frac{mv^2}{\rho}$, in which m is the mass of the moving load, v its velocity in feet per second, and ρ the radius of curvature of the beam. (See Mechanics.)

11. All these problems may be applied to beams fixed at one end, and P applied at the free end, or for a load uniformly distributed over the whole length, by using the equations under Cases I., II., and III.

According to equation (79) the deflection varies as the *cube of the length;* and inversely as the *breadth* and *cube of the depth*, and directly as the weight applied.

96. Barlow's Theory has not, to my knowledge, been applied to flexure, but it may be well to inquire what effect it would have. In the common theory, it is assumed that the total force is expended in elongating and compressing the fibres; but, according to Barlow's theory, a portion of the force is absorbed in drawing (so to speak) one fibre over the adjacent one; hence the deflection should be less by this theory than by the common one.

An experiment made by Mr. Hatcher, England, showed that it was less. (See Mosley's *Mechanics and Engineering*, p. 514.)

To find E by this theory, ϕ will represent a *fractional part of the strain* (not of the ultimate resistance).

Then $\phi \iint y\, dy\, dx$ is the moment of resistance to longitudinal shearing.

Hence we have

$$\Sigma Px - \phi \iint y\, dy\, dx = \frac{EI}{\rho}$$

Or for a rectangular beam supported at the ends, we have, by combining the general moments of equations (71) and (74) and using y positive upwards :—

$$(\tfrac{1}{2}P + \tfrac{1}{2}W - \tfrac{1}{2}wx)\, x - \tfrac{1}{4}bd^2\phi = EI\frac{d^2y}{dx^2} \quad - \; - \; - \; - \quad (81)$$

ϕ is *very small for* small deflections, but, whatever its value, we see that E found by this method will be less than that found by the common theory; and hence less than that given by the method in Article 7.

97. Case VII. Let the beam be fixed at one extremity, supported at the other, and have a weight, P, applied at any point.

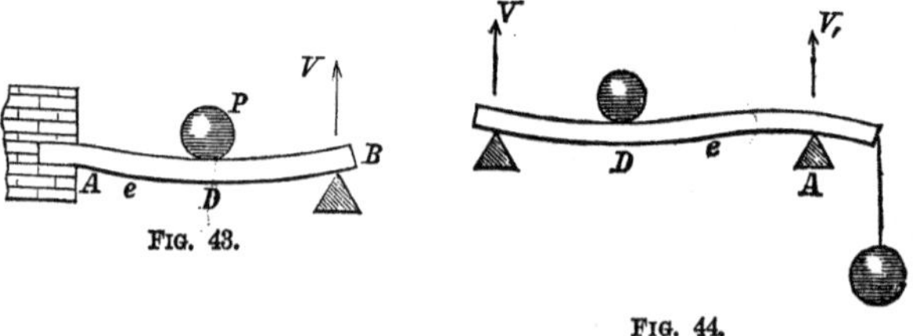

Fig. 43.

Fig. 44.

The beam may be fixed by being encased in a wall, Fig. 43, or by extending it over a support and suspending a weight on the extended part sufficient to make the beam horizontal over the support, Fig. 44; or by resting a beam whose length is $2l$ on three equidistant supports, and having two weights, each equal to P, resting upon it at equal distances from the central

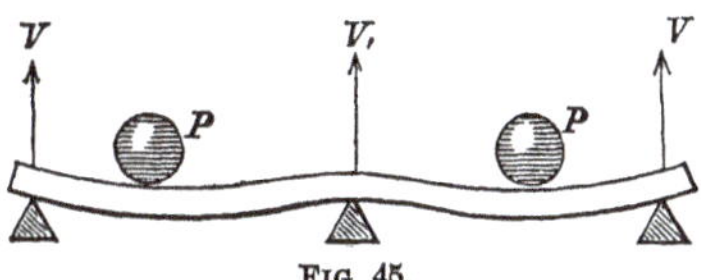

FIG. 45.

support, Fig. 45. In the latter case each half of the beam fulfils the condition of the case.

Let l = AB, Fig. 43, be the part considered,
V = the reaction of the support,
nl = AD = the abscissa of P, and
f = the deflection of the beam at D.

Take the origin at A, the fixed end. We may consider that the curve DB is caused by the reaction of V, while all the forces at the left of P hold the beam for V to produce its effect. Similarly the curve AD is produced by the reaction V and the weight P, while all the forces at the left of them hold the beam. In all cases we may consider that the applied forces on one side of the transverse section are in equilibrium with the resisting forces of tension and compression in the section. It is well also to observe that *the origin of moments is at the centre of the transverse section,* while the origin of coördinates may be at any point.

For the curve AD we have, observing that $\frac{dy}{dx} = 0$ for $x = 0$, and $y = 0$ for $x = 0$:—

$$\mathrm{EI}\frac{d^2y}{dx^2} = \mathrm{P}(nl - x) - \mathrm{V}(l - x), \qquad (82)$$

$$\mathrm{EI}\frac{dy}{dx} = \mathrm{P}\left(nlx - \frac{x^2}{2}\right) - \mathrm{V}\left(lx - \frac{x^2}{2}\right), \qquad (83)$$

$$\mathrm{EI}y = \mathrm{P}\left(\frac{nlx^2}{2} - \frac{x^3}{6}\right) - \mathrm{V}\left(\frac{lx^2}{2} - \frac{x^3}{6}\right). \qquad (84)$$

For the point D, we have, by making $x = nl$,

$$\frac{dy}{dx} = \tan i = [\tfrac{1}{2}n^2P - (n - \tfrac{1}{2}n^2)V] \frac{l^2}{EI}; \quad - \; - \; - \quad (85)$$

$$y = f = [\tfrac{1}{3}n^3P - (\tfrac{1}{2}n^2 - \tfrac{1}{6}n^3)V] \frac{l^3}{EI}. \quad - \; - \; - \; - \quad (86)$$

For the curve DB, observe that $\frac{dy}{dx} = \text{tang}\, i$ for $x = nl$, and $y = f$ for $x = nl$, using for their values (85) and (86) in determining the constants in the following equations, and we have:—

$$EI\frac{d^2y}{dx^2} = -V(l - x), \quad - \; - \; - \; - \; - \; - \; - \; - \; - \; - \quad (87)$$

$$EI\frac{dy}{dx} = \tfrac{1}{2}Pn^2l^2 - V\left(lx - \frac{x^2}{2}\right), \quad - \; - \; - \; - \; - \; - \quad (88)$$

$$EIy = (\tfrac{1}{2}x - \tfrac{1}{6}nl)Pn^2l^2 - V\left(\frac{lx^2}{2} - \frac{x^3}{6}\right). \quad - \; - \; - \; - \quad (89)$$

To find the reaction V, observe that $y = 0$, for $x = l$ in (89), and we obtain:—

$$0 = (3 - n)Pn^2l^3 - 2Vl^3;$$

$$\therefore V = \tfrac{1}{2}n^2(3 - n)P. \quad - \; - \; - \; - \quad (90)$$

By substituting this value of V in the preceding equations, they become completely determined. For the curve AD we shall have:—

$$EI\frac{d^2y}{dx^2} = P[nl - x - \tfrac{1}{2}n^2(3 - n)(l - x)]; \quad - \; - \quad (91)$$

$$\frac{dy}{dx} = \frac{P}{4EI}[4nlx - 2x^2 - n^2(3 - n)(2lx - x^2)], \quad - \quad (92)$$

$$y = \frac{P}{12EI}[6nlx^2 - 2x^3 - n^2(3 - n)(3lx^2 - x^3)]; - \quad (93)$$

and for the curve DB:—

$$EI\frac{d^2y}{dx^2} = -\tfrac{1}{2}Pn^2(3 - n)(l - x), \quad - \; - \; - \; - \; - \quad (94)$$

$$\frac{dy}{dx} = \frac{Pn^2}{4EI}[2l^2 - (3 - n)(2lx - x^2)], \quad - - - - - \quad (95)$$

$$y = \frac{Pn^2}{12EI}[(6x - 2nl)l^2 - (3lx^2 - x^3)(3-n)]. \quad - \quad (96)$$

The points of greatest strain in these curves are where the sum of the moments of applied forces is greatest, and this is greatest when the second members of (91) and (94) are greatest. Neither of these expressions have an algebraic maximum, and hence *we must find by inspection that value of x which will give the greatest value of the function within the limits of the problem.* Equation (91) has two such values, one for $x = 0$, the other for $x = nl$, and equation (94) has one such for $x = nl$, which value will reduce (91) and (94) to the same value.

Making $x = 0$ in (91) gives for the moment of maximum strain,

$$\Sigma Px = \tfrac{1}{2}Pl\,[2n - 3n^2 + n^3] \quad - - - - - - - \quad (97)$$

For the moment of strain at P, make $x = nl$, in (91) or (94), and we have

$$\Sigma Px = \tfrac{1}{2}\,Pln^2\,[-3 + 4n - n^2] \quad - - - - - - \quad (98)$$

To find where P must be applied so that the strain at the point of application shall be greater than if applied at any other point, we must find the maximum of (98):—

$$\therefore D_n = 0 = -6n + 12n^2 - 4n^3 \quad - - - - - \quad (99)$$

$$\therefore n = 0.634 + \quad - - - - - - - - - - - \quad (100)$$

or the force must be applied at more than $\frac{63}{100}$ of the length of the beam from the fixed end. This value of n in (98) gives,

$$\Sigma Px = Pl \times 0.174$$

Equation (99) has two values of n, but the other is not within the limits of the problem.

The position of the weight, which will give a maximum strain at the fixed end, is found by making (97) a maximum. Proceeding in the usual way, we find:—

$$n = 1 \pm \tfrac{1}{3}\sqrt{3} = 0.422 + \quad - - - - - - - - \quad (101)$$

which in (97) gives, $\Sigma Px = Pl \times 0.181 \quad - - - - - \quad (102)$

and in (98) $\quad \Sigma Px = Pl \times 0.131 +$

To find where P must be applied so that the strain at the

point of application will equal the strain at the fixed end, make equations (97) and (98) equal to each other, and find n. This gives,

$$n = \begin{cases} 1. \\ 3.4141+; \\ 0.5858+. \end{cases} \quad (103)$$

But $n = 0.5858 +$ is the only practical value.

To find where P must be applied so that the curve at that point shall be horizontal, make $\frac{dy}{dx} = 0$, and $x = nl$ in (95).

$$\text{This gives } n = \begin{cases} 1. \\ 3.4141 \\ 0.5858 \end{cases}$$

which are the same as the preceding values of n. To find the corresponding deflection, make $x = nl$, and $n = 0.5858 +$, in (93), and we find

$$\Delta = 0.0098 \frac{Pl^3}{EI} \quad (104)$$

$$\left.\begin{array}{l} \text{For } n < 0.5858, \text{ tang } i \text{ is } + \\ \quad n > 0.5858, \text{ tang } i \text{ is } - \\ \quad n = 0.5858, \text{ tang } i \text{ is } 0 \end{array}\right\} \quad (105)$$

To find the maximum deflection when $n = 0.634$, make $\frac{dy}{dx} = 0$ in (92) or (95), according as the greater deflection is to the right or left of P. But, according to (105), it belongs to the curve AD; hence use (92). Making $n = 0.634$ in (92), placing it equal zero, and solving gives,

$$x = 0.6045l;$$

which in (93) gives,

$$y = \Delta = 0.00957 \frac{Pl^3}{EI}. \quad (106)$$

To find where P must be applied so as to give an absolute maximum deflection; first find the abscissa of the point of maximum deflection, when P is applied at any point by making $\frac{dy}{dx} = 0$ in (92), and thus find

$$x = \frac{2(3-n)n^2 - 4n}{(3-n)n^2 - 2} l; \quad (107)$$

which, substituted in (93) gives the corresponding maximum

deflection. Then find that value of n which will make the expression a maximum.

The point of contra-flexure in the curve AD is found by making $\frac{d^2y}{dx^2}=0$ in (91) (see Dif. Cal.) which gives,

$$x=\frac{3ln^2-n^3l-2nl}{3n^2-n^3-2}$$

If $n=\frac{1}{2}$, $x=\frac{3}{11}l$.

The second member of (91) is the moment of applied forces, and as it is naught at the point of contra-flexure, it follows that at that point there is no bending stress, and hence no elongation or compression of the fibres, but only a transverse shearing stress, the value of which is determined in Article 153.

If a beam rests upon three horizontal equidistant supports, and two weights, each equal P, are placed upon it, one on each side of the central support and equidistant from it, it fulfills the condition of a beam fixed at one end and supported at the other, as before stated, and the amount which each support will sustain for incipient flexure may easily be found from the preceding equations.

The three supports will sustain 2P, and the end supports each sustain $V=\frac{1}{2}n^2(3-n)P$. (See Eq. (90).)

Hence, the central support sustains

$$V'=2P-n^2(3-n)P.$$

If $n=\frac{1}{2}$, $V=\frac{5}{16}P$, and $V'=\frac{22}{16}P$.

98. CASE VIII.—LET THE BEAM BE FIXED AT ONE END, SUPPORTED AT THE OTHER, AND UNIFORMLY LOADED OVER ITS WHOLE LENGTH.

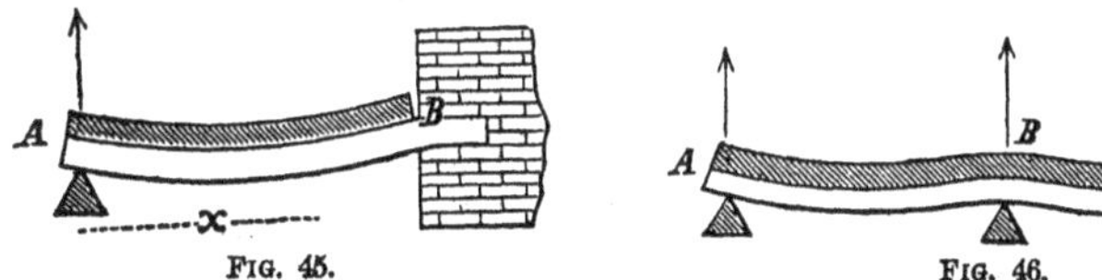

FIG. 45. FIG. 46.

Take the origin at A, Figs. 45 and 46, and the notation the same as in the preceding cases, then equation (50) becomes

$$EI\frac{d^2y}{dx^2}=\frac{1}{2}wx^2-Vx \quad (108)$$

Integrating gives $\frac{dy}{dx} = \frac{w}{6EI}(x^3 - l^3) + \frac{V}{2EI}(l^2 - x^2)$, (109)

and $y = \frac{w}{24EI}(x^4 - 4l^3 x) + \frac{V}{6EI}(3l^2 x - x^3)$; - (110)

in which $\frac{dy}{dx} = 0$ for $x = l$, and $y = 0$ for $x = 0$.

If $V = 0$, these equations become the same as those under CASE II.

In equation (110) y is also zero, for $x = l$; for which values we have $V = \frac{3}{8} W = \frac{3}{8} wl$ - - - - - - - - - (111)

This value substituted in equations (108), (109), and (110) gives:—

$$EI \frac{d^2y}{dx^2} = \tfrac{1}{8} wx (4x - 3l); \text{ - - - - - - - - (112)}$$

$$\frac{dy}{dx} = \frac{w}{48EI}(8x^3 - 9lx^2 + l^3); \text{ - - - - - - - (113)}$$

$$y = \frac{w}{48EI}(2x^4 - 3lx^3 + l^3 x). \text{ - - - - - - - (114)}$$

The point of maximum deflection is found by placing equation (113) equal zero and solving for x. This gives

$$x = \frac{1 \pm \sqrt{33}}{16} l = 0.4215l;$$

and this in (114) gives

$$y = \Delta = 0.0054 \frac{W l^3}{EI} \text{ - - - - - - - (115)}$$

There are two maxima strains; one for $x = l$; the other for $x = \frac{3}{8} l$. The former in (112) gives

$$\Sigma Px = \tfrac{1}{8} wl^2 = \tfrac{1}{8} Wl, \text{ - - - - - - - (116)}$$

and the latter gives

$$\Sigma Px = -\tfrac{9}{128} Wl = -\tfrac{1}{14} Wl \text{ nearly.}$$

The point of contra-flexure is found from equation (112) to be at $x = \frac{3}{4} l$, at which point the longitudinal strains are zero, and there is only transverse shearing. (See Article 84.)

If the beam is supported by three props, which are in the same horizontal, Fig. 46, then each part is subjected to the same conditions as the single beam in Fig. 45. Hence, if W is the load on half the beam, each of the end props will sustain $V = \frac{3}{8}W$, (Eq. (111)), and the middle prop will sustain $2W - \frac{6}{8}W = \frac{5}{4}W$.

Such are the teachings of the "common theory." But the mathematical conditions here imposed are never realized. It is impossible to maintain the props exactly in the same horizontal. As they are elastic they will be compressed, and as the central one will be most compressed, the tendency will be to relieve the strain on it and throw a greater strain upon the end supports. If the supports be maintained in the same horizontal, the results above deduced will be practically true for *very small* deflections, but will be somewhat modified as the strains approach the breaking limit.

99. CASE IX.—LET THE BEAM BE FIXED AT BOTH ENDS AND A WEIGHT REST UPON IT AT ANY POINT.

To simplify the case, suppose that the weight rests at the middle of the length.

Let the beam be extended over one support and a weight, P_1 rest at C, sufficient to make the curve horizontal over the support A. We have $V = P_1 + \frac{1}{2}P$.

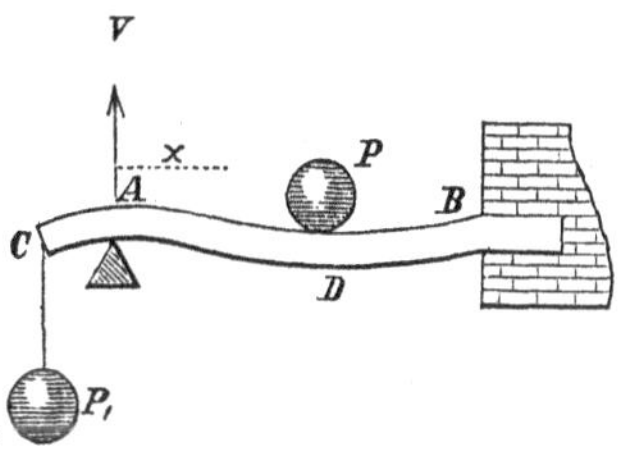

FIG. 47.

Let $AC = ql$.

Then for the curve AD we have,

$$EI\frac{d^2y}{dx^2} = P_1(ql + x) - Vx = P_1 ql - \tfrac{1}{2}Px$$

$$\therefore EI\frac{dy}{dx} = P_1 qlx - \tfrac{1}{4}Px^2 + (C_1 = 0).$$

To find P_1 observe that $\frac{dy}{dx} = 0$ for $x = \frac{1}{2}l$;

$$\therefore 0 = \tfrac{1}{2}P_1 ql^2 - \tfrac{1}{16}Pl^2;\ \therefore P_1 q = \tfrac{1}{8}P.$$

This reduces the preceding equations to the following:—

$$EI\frac{d^2y}{dx^2} = \tfrac{1}{8}P(l - 4x) \quad - \; - \; - \; - \quad (117)$$

$$\frac{dy}{dx} = \frac{P}{8EI}(lx - 2x^2) \quad - \; - \; - \; - \; - \quad (118)$$

and by integrating again, we find:—

$$y = \frac{P}{48EI}(3lx^2 - 4x^3) \quad - \; - \; - \; - \quad (119)$$

For $x = \frac{1}{2}l$ in (119), $y = \Delta = \frac{Pl^3}{192EI}$ - - - (120)

There is no algebraic maximum of the moment of strain as given in the second member of equation (117), but inspection shows that within the limits of the problem the moment is greatest for $x = 0$ or $x = \frac{1}{2}l$. These in (117) give the same value, with contrary signs; hence the moment of greatest strain is

$$\Sigma Px = \pm \tfrac{1}{8}Pl \quad \text{- - - - - - - - - -} \quad (121)$$

The moment is zero for $x = \frac{1}{4}l$.

100. CASE X.—LET THE BEAM BE FIXED AT BOTH ENDS AND A LOAD UNIFORMLY DISTRIBUTED OVER ITS WHOLE LENGTH.

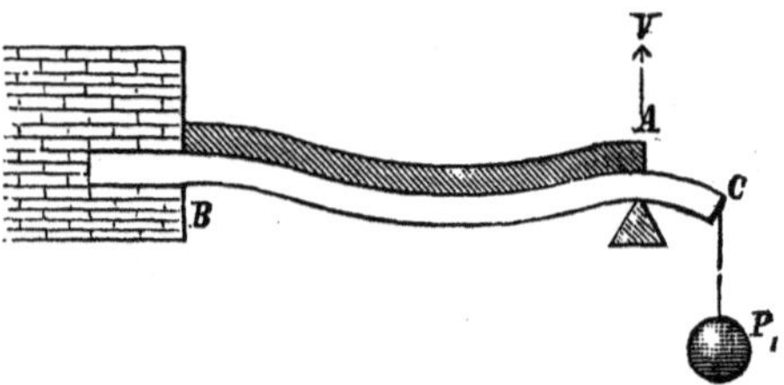

FIG. 48.

The notation being the same as before used, we have

$$V = P_1 + \tfrac{1}{2}wl.$$

Let $ql = AC$.

The equation of moments is

$$EI\frac{d^2y}{dx^2} = \tfrac{1}{2}wx^2 - Vx + P_1(ql + x)$$

$$= \tfrac{1}{2}wx^2 - \tfrac{1}{2}wlx + P_1ql.$$

Integrating, and observing that $\frac{dy}{dx} = 0$ for $x = 0$; also $y = 0$ for $x = 0$, and we have

$$EI\frac{dy}{dx} = \tfrac{1}{6}wx^3 - \tfrac{1}{4}wlx^2 + P_1qlx$$

$$EIy = \tfrac{1}{24}wx^4 - \tfrac{1}{12}wlx^3 + \tfrac{1}{2}P_1qlx^2.$$

But $\frac{dy}{dx} = 0$ for $x = l$; also $y = 0$ for $x = l$;

$$\therefore P_1 = \frac{1}{12}\frac{wl}{q} = \frac{W}{12q}$$

which substituted in the previous equations give:—

$$\mathrm{EI}\frac{d^2y}{dx^2}=\frac{\mathrm{W}}{12l}\left[l^2-6x(l-x)\right] \quad - \quad - \quad - \quad (122)$$

$$\frac{dy}{dx}=\frac{\mathrm{W}}{12\mathrm{EI}l}\left[(l-x)(l-2x)x\right] \quad - \quad - \quad (123)$$

$$y=\frac{\mathrm{W}}{24\mathrm{EI}l}\left[(l-x)^2x^2\right] \quad - \quad - \quad - \quad - \quad - \quad (124)$$

For $x=\frac{1}{2}l$ in (124), $y=\triangle=\frac{1}{384}\frac{\mathrm{W}l^3}{\mathrm{EI}}$ - - - - - - (125)

Making $\frac{d^2y}{dx^2}=0$ we find for the points of contra-flexure

$$x=\begin{cases}0.7887l\\0.2113l\end{cases}$$

at which points there is no longitudinal strain, but a transverse shearing strain. (See Article 84.)

The maximum moments are for $x=0$ and $x=\frac{1}{2}l$.
For $x=0$, the second member of Eq. (122) gives $\frac{1}{12}\mathrm{W}l$. (126)
For $x=\frac{1}{2}l$, " " " $\frac{1}{24}\mathrm{W}l$.

Hence the greatest strain is over the support, at which point it is twice as great as at the middle. If W = P, we see that the strain over the support is $\frac{2}{3}$ as great in this case as in the former.

101. RESULTS COLLECTED.

No. of the Case.	Condition of the Beam.	How Loaded.	General Moment of Flexure.	Maximum Moment of Stress.	Relative Moment.	Relative Max. Deflection or Coefficient of $\frac{l^3}{EI}$
I.	Fixed at one end.	Load at free end.	Px. *Eq.* (53).	Pl.	24	$\frac{1}{3}P$. *Eq.* (57).
II.		Uniform load.	$\frac{1}{2}wx^2$. *Eq.* (58).	$\frac{1}{2}Wl$.	12	$\frac{1}{8}W$. *Eq.* (61).
IV.	Supported at the ends.	At the middle.	$\frac{1}{2}Px$. *Eq.* (71).	$\frac{1}{4}Pl$.	6	$\frac{1}{48}P$. *Eq.* (73).
V.		Uniform.	$\frac{1}{2}w(lx-x^2)$. *Eq.* (74).	$\frac{1}{8}Wl$.	3	$\frac{5}{384}W$. *Eq.* (76).
VII.	Fixed at one end and supported at the other.	At 0.634l from fixed end. *Eq.* (100).	For AD *Eq.* (91). For DB *Eq.* (94).	$\frac{3}{8}(2\sqrt{3}-3)Pl$.	4+	$\frac{P}{102}$ nearly. *Eq.* (104).
VIII.		Uniform.	$\frac{1}{8}w(4x^2-3lx)$. *Eq.* (112).	$\frac{1}{8}Wl$. *Eq.* (116).	3	$\frac{W}{185}$ nearly. *Eq.* (115).
IX.	Fixed at both ends.	At the middle.	$\frac{1}{8}P(l-4x)$. *Eq.* (117).	$\frac{1}{8}Pl$. *Eq.* (121).	3	$\frac{P}{192}$. *Eq.* (120).
X.		Uniform.	$\frac{W}{12l}(l^2-6lx+6x^2)$. *Eq.* (122).	$\frac{1}{12}Wl$. *Eq.* (126).	2	$\frac{W}{384}$. *Eq.* (125).

102. REMARKS.—It will be seen that the greatest strains in the 1st and 2d cases are as 2 to 1; and the same ratio holds in the 4th and 5th cases; but in the 9th and 10th the ratio is as 3 to 2. This is peculiar, and further remarks are made upon it in Article 119. The maximum strains in Cases VII. and VIII. do not occur at the points of maximum deflection.

Although the moment in the 1st case is to that in the 2d as 2 to 1, yet the deflections are as 8 to 3; and in the 4th and 5th cases the deflections are as 8 to 5.

A comparison of Cases IV. and IX. shows the advantage of fixing the ends of the beam. The same remark applies to Cases V. and X. In the former cases the strain is only one-half as great when the beam is fixed at the ends as when it is supported, and in the latter two-thirds as great.

Other interesting results may be seen by examining the table.

103. MODIFICATION OF THE FORMULAS FOR DEFLECTION.—It will be observed that the general form of the expression for the maximum deflection of rectangular beams is

$$\triangle = \text{constant} \times \frac{Pl^3}{Ebd^3}$$

Prof. W. A. Norton, of New Haven, Ct., has made experiments to test the correctness of this expression. (See *Van Nostrand's Eclectic Engineering Magazine*, vol. 3, page 70.) According to his experiments, *for beams supported at their ends and loaded at the middle*, the expression should be

$$\triangle = \frac{Pl^3}{4Ebd^3} + C\frac{Pl}{bd} \quad - \; - \; - \; - \; - \; - \quad (126a)$$

For the pine sticks which he used he found the mean value of C to be

$$C = 0.0000094.$$

A consideration of transverse shearing stress, in combination with the stretching and compressing of the fibres, leads to an expression of this form. For, as we have before seen, the strain is evenly distributed over the whole transverse section, and hence the deflection will vary inversely as the area, or as bd; it is also uniform over its whole length, and equal $\frac{1}{2}$P (see Example 2, Article 84); and hence the amount of deflection will vary as $\frac{1}{2}$P; and the total deflection at the middle will evidently vary as the length; or, in this case, as $\frac{1}{2}l$. Hence, the total deflection due to transverse shearing is, $C\frac{\frac{1}{2}P.\frac{1}{2}l}{bd} = \frac{C}{4}\frac{Pl}{bd}$, which is the same form as that given by Professor Norton. The same form of expression is also reached, in a more circuitous way, by Weisbach, in his *Mechanics of Engineering*, 4th edition, vol. 1, page 522 of the recent American edition.

In Professor Norton's formula $\frac{1}{4}$ C is the reciprocal of the coefficient of elasticity to transverse shearing of *white pine;* hence the coefficient is 425,531 pounds. The mean value of E in the above experiments was found to be 1,427,965 lbs.; and hence, in this case, the reciprocal of $\frac{1}{4}$C is a little more than $3\frac{1}{3}$ of E. Weisbach, in the reference above given, says: "The coefficient of elasticity for transverse shearing is generally assumed to be equal to $\frac{3}{8}$E."

If the load is uniformly distributed over the whole length, the shearing stress on any section, distant x from the end, is $\frac{1}{2}wl - wx$. (See Example 1, Article 84.) Hence the deflection for an element of length of a rectangular beam due to this cause, is

$$C\frac{(\frac{1}{2}wl - wx)dx}{bd},$$

and for a distance x this becomes by simple integration,

$$C\frac{w}{2bd}(l - x)x,$$

and for half the length, make $x = \frac{1}{2}l$, and the expression becomes

$$C\frac{wl^2}{8bd} = C\frac{Wl}{8bd};$$

from which we see that the same load, distributed uniformly over the whole length, produces half as much deflection due to *transverse shearing* as the same load concentrated at the middle.

Equation (76) when corrected for this effect becomes

$$\triangle = Wl\left[\frac{5l^2}{384EI} + C\frac{1}{8bd}\right]$$

$$\text{or, } \triangle = \frac{Wl}{8bd}\left[\frac{5l^2}{4Ed^2} + C\right]$$

from which we see that if the depth be constant the deflection due to transverse shearing will be more apparent compared with that due to the other cause, as the piece is shorter. If the piece is very long, the effect due to C is comparatively small. If $\frac{4}{C} = \frac{1}{3}E$, as assumed by Weisbach, the deflection becomes

$$\triangle = \frac{Wl}{8Ebd}\left[\frac{5}{4}\frac{l^2}{d^2}+12\right]$$

If $l = d$ the quantity in [ ] becomes $\frac{5}{4}+12$; or the effect due to C is $\frac{48}{5}$ of that due to E.

If $l = 20d$, the effect due to C is $\frac{1}{42}$ nearly of that due to E.

104. ADDITIONAL PROBLEMS WHICH ARE PURPOSELY LEFT UNSOLVED.

1. Suppose that a beam is supported at its extremities, and has two forces at any point between. In this case the curve between the support and the nearest force will have one equation; the curve between the forces another; and the remaining part a third.

2. In the preceding case, if the forces are equal and equidistant from the supports, the curve between the forces will be the arc of a circle.

3. Suppose that the beam is uniformly loaded and rests on four supports.

4. Suppose that the beam is supported at its extremities and has a load uniformly increasing from one support to the other.

5. Suppose that the beam is uniformly loaded over any portion of its length.

6. Suppose that it has forces applied at various points.

These problems will suggest many others.

7. Suppose that a beam is supported at several points, and loaded uniformly over its whole length.

Let W = the weight between each pair of supports,

V_1, V_2, V_3, &c., be the reactions of the supports, counting from one end,

and let the distances between the supports be equal.

Then we have :—

No. of Supports.	V_1	V_2	V_3	V_4	V_5	V_6	V_7	V_8	V_9	V_{10}
2	$\frac{1}{2}$W	$\frac{1}{2}$W	Frac	tional	parts	of W.				
3	$\frac{3}{8}$	$\frac{10}{8}$	$\frac{3}{8}$							
4	$\frac{4}{10}$	$\frac{11}{10}$	$\frac{11}{10}$	$\frac{4}{10}$						
5	$\frac{11}{28}$	$\frac{32}{28}$	$\frac{26}{28}$	$\frac{32}{28}$	$\frac{11}{28}$					
6	$\frac{15}{38}$	$\frac{43}{38}$	$\frac{37}{38}$	$\frac{37}{38}$	$\frac{43}{38}$	$\frac{15}{38}$				
7	$\frac{41}{104}$	$\frac{118}{104}$	$\frac{108}{104}$	$\frac{106}{104}$	$\frac{108}{104}$	$\frac{118}{104}$	$\frac{41}{104}$			
8	$\frac{56}{142}$	$\frac{161}{142}$	$\frac{137}{142}$	$\frac{143}{142}$	$\frac{143}{142}$	$\frac{137}{142}$	$\frac{161}{142}$	$\frac{56}{142}$		
9	$\frac{153}{388}$	$\frac{440}{388}$	$\frac{374}{388}$	$\frac{392}{388}$	$\frac{386}{388}$	$\frac{392}{388}$	$\frac{374}{388}$	$\frac{440}{388}$	$\frac{153}{388}$	
10	$\frac{209}{530}$	$\frac{601}{530}$	$\frac{511}{530}$	$\frac{535}{530}$	$\frac{529}{530}$	$\frac{529}{530}$	$\frac{535}{530}$	$\frac{511}{530}$	$\frac{601}{530}$	$\frac{209}{530}$

If the beams and props were perfectly rigid, all but the end ones would sustain W, and the end ones each $\frac{1}{2}$ W.

It may be shown that, for any number of equidistant props, the inclination at the end may be found from the equation

$$\text{tang}\, i = (4V_1 - W)\frac{l^2}{24EI}$$

which for 10 props becomes

$$\text{tang}\, i = \frac{153}{265}\,\frac{Wl^2}{24EI};$$

and the maximum deflection for any number of props is

$$\triangle = (24V_1 - 7W)\frac{l^3}{384EI}$$

105. BEAMS OF VARIABLE SECTIONS.

For these I is variable, and its value must be substituted in equation (50) before the integration can be performed. As an example, let the beam be fixed at one extremity, and a weight, P, be suspended at the free extremity, Fig. 50. Let the breadth be constant, and the longitudinal vertical sections be a parabola. Then all the transverse sections will be rectangles.

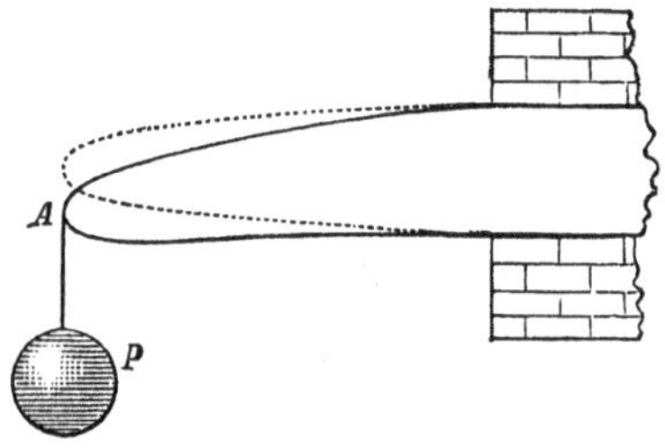

FIG. 50.

Let $l =$ the length,
$b =$ the breadth, and
$d =$ the depth at the fixed extremity.

If y is the whole variable depth at any point, we have, from the equation of the parabola,

$(\frac{1}{2}y)^2 = px$, or $\frac{1}{4}d^2 = pl$, $\therefore p = \frac{d^2}{4l}$, in which p is the parameter of the parabola.

$$\therefore y^2 = \frac{d^2x}{l}, \quad - \; - \; - \; - \; - \; - \; - \; - \; - \; - \quad (127)$$

From equation (51) we have

$I = \frac{1}{12}by^3$, in which substitute y, from equation (127), and we have $I = \frac{1}{12}\frac{bd^3}{l^{\frac{3}{2}}}x^{\frac{3}{2}}$ $\quad - \; - \; - \; - \; - \; - \; - \; - \; - \; - \quad (128)$

The equation of moments is, see equation (50),

$EI\frac{d^2y}{dx^2} = Px$, in which substitute I, from equation (128), and we have

$$\frac{d^2y}{dx^2} = \frac{12Pl^{\frac{3}{2}}}{Ebd^3}x^{-\frac{1}{2}}$$

Multiply by the dx and integrate, observing that $\frac{dy}{dx} = 0$ for $x = l$ and we have

$$\frac{dy}{dx} = \frac{24Pl^{\frac{3}{2}}}{Ebd^3}\left(x^{\frac{1}{2}} - l^{\frac{1}{2}}\right)$$

Integrating again gives

$$y = \frac{8Pl^{\frac{3}{2}}}{Ebd^3}(2x^{\frac{3}{2}} - 3l^{\frac{1}{2}}x)$$

$$y \text{ is zero for } x = 0.$$
$$y = \Delta \quad \text{for } x = l;$$

$$\therefore \Delta = -\frac{8Pl^3}{Ebd^3} \quad \text{- - - - - - - - - - (129)}$$

If, in equation (57), we substitute $I = \frac{1}{12}bd^3$ (Eq. (51)), it becomes

$$\Delta = -\frac{4Pl^3}{Ebd^3}$$

which is one-half that of (129); hence the deflection of a prismatic beam is one-half that of a parabolic beam of the same length, breadth, and greatest depth, when fixed at one end and free at the other, and has the same weight suspended at the free end.

In a similar manner the equation of the curve may be found for any other form of beam, if the law of increase or decrease of section is known. Several examples may be made of beams of uniform strength, which will be given in Chapter VII.

106. BEAMS SUBJECTED TO OBLIQUE STRAINS.—*Let the beam be prismatic, fixed at one end, and support a weight,* P, *at the free end; the beam being so inclined that the direction of the force shall make an obtuse angle with the axis of the beam, as in* Fig. 51.

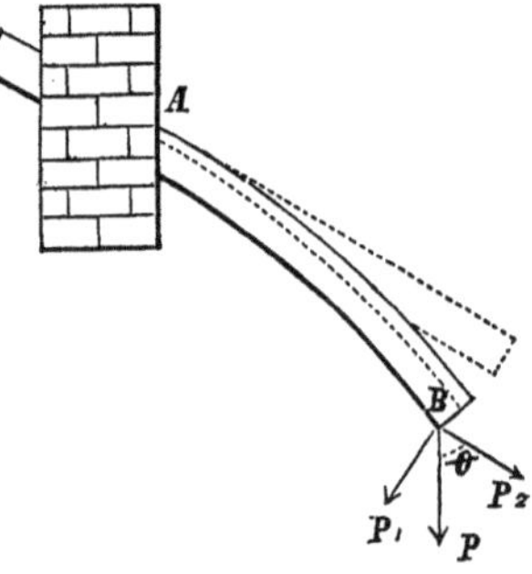

FIG. 51.

Let $P_1 = P \sin \theta$ = component of P perpendicular to the axis of the beam, and

$P_2 = P \cos \theta$ = component parallel to the axis of the beam.

Take the origin at the free end, the axis of x being parallel to the axis of the beam, and y perpendicular to it.

Then equation (50) becomes

$$EI\frac{d^2y}{dx^2} = -P_1x + P_2y$$

$$\text{or, } \frac{d^2y}{dx^2} = -p^2x + q^2y \quad \text{- - - - - - - - - - - (130)}$$

in which $p^2 = \frac{P_1}{EI}$; and $q^2 = \frac{P_2}{EI}$. The complete integral of (130) is (see Appendix III.)

$$y = C_1 e^{qx} + C_2 e^{-qx} + \frac{p^2}{q^2}x$$

The conditions of the problem give

$\frac{dy}{dx} = 0$ for $x = l$;

$y = 0$ for $x = 0$; and these combined with the preceding equation give :—

$$0 = q\left(C_1 e^{ql} + C_2 e^{-ql} \right) + \frac{p^2}{q^2} l;$$

$$0 = C_1 + C_2;$$

From which C_1 and C_2 may be found, and the equation becomes completely known.

We also have $y = \Delta$ for $x = l$;

$$\therefore \Delta = C_1 e^{ql} + C_2 e^{-ql} + \frac{p^2}{q^2} l;$$

Next, suppose that the force makes an acute angle with the axis of the beam, as in Fig. 52.

For the sake of variety, take the origin at A, the fixed end, x, still coinciding with the axis of the beam before flexure. Using the same notation as in the preceding and other cases, we have

$$\frac{d^2y}{dx^2} = p^2(l - x) + q^2(\Delta - y) \quad \text{- - - - - - - - -} \quad (131)$$

The complete integral is

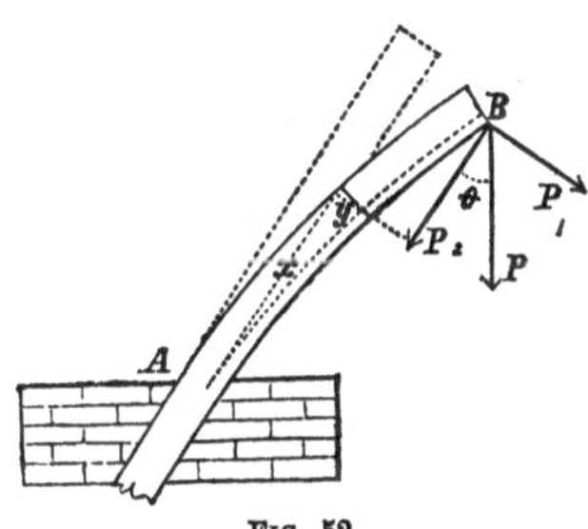

Fig. 52.

$$\Delta - y = \text{A} \sin q(x + \text{B}) - \frac{p^2}{q^2}(l - x) \quad (132)$$

in which A and B are arbitrary constants.

From the problem we have

$y = 0$ for $x = 0$;

$\frac{dy}{dx} = 0$ for $x = 0$; and

$y = \Delta$ for $x = l$;

by means of which the equation becomes completely known.

From these examples we see how easily the problem is complicated. One difficulty in applying these cases in practice is in determining the value of I. Before it can be determined, the position of the neutral axis must be known. According to Article 78, 3d case, it appears that the neutral axis does not coincide with the axis of the beam. Indeed, according to the same article, it is not parallel to the axis, and hence I is variable, and the equations above are only a secondary approximation; the first approximation being made in establishing equation (50), and the next one in assuming I constant. All writers, within the author's acquaintance, who have investigated this and similar cases, beginning with Navier, have assumed that I is constant, and that the neutral axis coincides with the axis of the beam. These assumptions may be admissible in any practical case were extreme accuracy is not desired. Many other practical examples might be given, the solutions of which are more difficult than the preceding; but enough have been given to illustrate the methods.

107. FLEXURE OF COLUMNS.—If a weight rests upon the axis of a perfectly symmetrical and homogeneous column, we see no reason why it should bend it; but in practice we know that it will bend, however symmetrical and homogeneous it may be, and however carefully the weight may be placed upon it. If the weight be small, the deflection may not be visible to the unaided eye. If the weight is not so heavy as to crush the column, an equilibrium will be established between the weight and the elastic resistances within the beam. Let the column rest upon a horizontal plane, and the weight P on the upper end be vertically over the lower end. Take the origin of coördinates at the lower end of the column, Fig. 53, x being vertical, and y horizontal. They must be so taken here, because x was assumed to coincide with the axis of the beam when equation (50) was established. Then y being the ordinate to any point of the axis of the column after flexure, the moment of P is Py, which is negative in reference to the moment of resisting forces, because the curve is concave to the axis of x, in which case the ordinate and second differential coefficient must have contrary signs (Dif. Cal.). Hence we have,

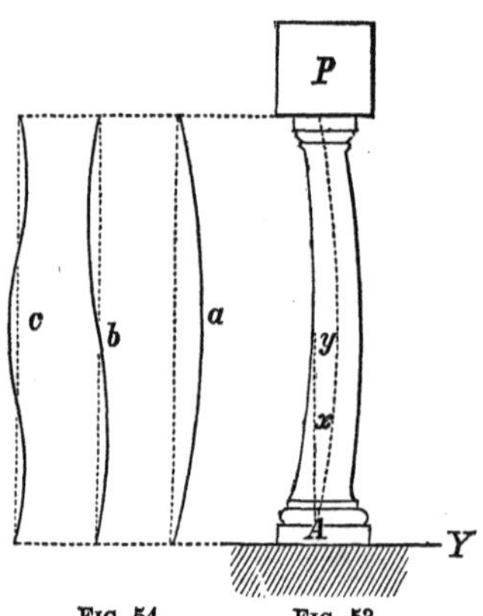

FIG. 54. FIG. 53.

$$EI\frac{d^2y}{dx^2} = -Py \quad - - - - - - - - - - \quad (133)$$

Multiply by dy and integrate (observing that dx is constant), and find

$$\frac{dy^2}{dx^2} = -\frac{P}{EI} y^2 + C_1.$$

But $\frac{dy}{dx} = 0$ for $y = \Delta =$ the maximum deflection. These values in the preceding equation give $C_1 = \frac{P\Delta^2}{EI}$, which being substituted in the same equation and reduced gives

$$dx = \sqrt{\frac{EI}{P}} \times \frac{dy}{\sqrt{\Delta^2 - y^2}}$$

$$\therefore x = \sqrt{\frac{EI}{P}} \sin^{-1} \frac{y}{\Delta} + C_2.$$

But $y = 0$ for $x = 0 \therefore C_2 = 0$. Hence the preceding gives

$$y = \Delta \sin \sqrt{\frac{P}{EI}} x \quad - \; - \; - \; - \; - \; - \; - \; - \quad (134)$$

But $y = 0$ for $x = l$. Therefore, if n is an integer, these values reduce (134) to

$$\sqrt{\frac{P}{EI}} \times l = n\pi;$$

$$\therefore P = EI \frac{n^2\pi^2}{l^2} \quad - \; - \; - \; - \; - \; - \; - \quad (135)$$

This value of P reduces (134) to

$$y = \Delta \sin n\pi \frac{x}{l}$$

which is the equation of the curve. It is dependent only upon the length of the column and the maximum deflection. If $n = 1$, the curve is represented by a, Fig. 54; if $n = 2$, by b; if $n = 3$, by c.

If $n = 1$, equation (135) becomes

$$P = \frac{\pi^2}{l^2} EI \quad - \; - \; - \; - \; - \; - \; - \; - \quad (136)$$

which is the formula to be used in practice. We see that the resistance is independent of the deflection. If the column is cylindrical, $I = \frac{1}{4}\pi r^4$ (see equation (52));

$$\therefore P = \frac{\pi^2 E}{4} \times \frac{r^4}{l^2} \quad - - - - - - - - \quad (137)$$

hence the resistance varies as the fourth power of the radius (or diameter), and inversely as the square of the length. If the column is square, $I = \frac{1}{12} b^4$ (equation (51)),

$$\therefore P = \frac{\pi^2 E}{12} \times \frac{b^4}{l^2} \quad - - - - - - - - \quad (138)$$

These formulas, according to Navier * and Weisbach,† should be used only when the length is 20 times the diameter for cylindrical columns, or 20 times the least thickness for rectangular columns; and Navier says that for safety only $\frac{1}{10}$ of the calculated weight should be used in case of wood, and $\frac{1}{4}$ to $\frac{1}{6}$ in case of iron; but Weisbach says they should have a twenty-fold security.

Examples 1.—What must be the diameter of a cast-iron column, whose length is 12 feet, to sustain a weight of 30 tons (of 2,000 lbs. each); $E = 16,000,000$ lbs.; and factor of safety $\frac{1}{20}$. Ans. $d = 7.52$ in.

2. If the column be square and the data the same as in the preceding example, equation (138) gives

$$b = \sqrt[4]{\frac{12 \times 60,000 \times (12 \times 12)^2 \times 20}{(3{\cdot}1416)^2 \times 16,000,000}} = 6.6 \text{ inches.}$$

In the analysis of this problem I have followed the method of Navier; but as it is well known that the results are not relied upon by practical men, I have given only one case. For other cases see Appendix III. There are some reasons for the failure of the theory which are quite evident, but it is not easy to remedy them; and for this reason the empyrical formulas of Article 52 are much more satisfactory. It will be observed that the *law* of strength, as given in the formulas in that article, are the same as those given in equations (137) and (138) for wooden columns, and nearly the same as for iron ones. The chief difference is in the coefficients, or constant factors. In the analysis it was assumed that the neutral axis coincides with the axis of the beam, but it is possible for the whole column to be compressed, although much more on the concave than on the convex side, in which

* Navier, Résumé des Leçons, 1839, p. 204.

† Weisbach's Mechanics and Engineering. Vol. 1, p. 219. 1st Am. ed.

case the neutral axis would be ideal, having its position entirely outside the beam on the convex side. In this case, if the ideal axis is parallel to the axis of the beam, it does not affect the form of equation (136), but it does affect the value of I; and hence the values of equations (137) and (138). The problem of the flexure of columns is then more interesting as an analytical one than profitable as a practical one.

GRAPHICAL METHOD.

108. THE GRAPHICAL METHOD consists in representing quantities by geometrical magnitudes, and reasoning upon them, with or without the aid of algebraic symbols. This method has some advantages over purely analytical processes; for by it many problems which involve the spirit of the Differential and Integral Calculus may be solved without a knowledge of the processes used in those branches of mathematics; and in some of the more elementary problems, in which the spirit of the Calculus is not involved, the quantities may be directly presented to the eye, and hence the solutions may be more easily retained. It is distinguished, in this connection, from pure geometry by being applied to problems which involve mechanical principles, and to use it profitably in such cases requires a knowledge of the elementary principles of mechanics as well as of geometry.

But graphical methods are generally special, and often require peculiar treatment and much skill in their management. It is not so powerful a mode of analysis as the analytical one, and those who have sufficient knowledge of mathematics to use the latter will rarely resort to the former, unless it be to illustrate a principle or demonstrate a problem for those who cannot use the higher mathematics. A few examples will now be given to illustrate the method.

109. GENERAL PROBLEM OF THE DEFLECTION OF BEAMS.—*To find the total deflection of a prismatic beam which is bent by a force acting normal to the axis of the beam* without the aid of the Calculus.

Let a beam, AB, Fig. 55, be bent by a force, P, in which case the fibres on the convex side will be elongated, and those on the concave side will be compressed. Let AB be the neutral axis. Take two sections normal to the neutral axis at L and N, which are *indefinitely near each other.* These, if prolonged, will meet at some point as O. Draw KN parallel to LO. Then will $ke, = \lambda$, be the distance between KN and EN at k, and is the elongation of the fibre at k. Let $eN = y$, then from the similar triangles kNe and LON we have

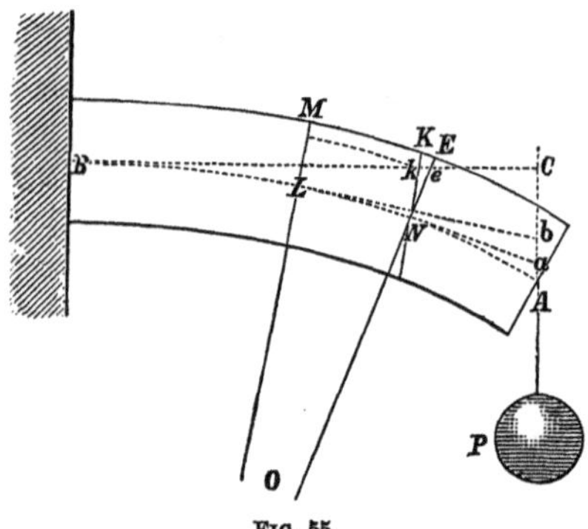

FIG. 55.

$$ON : Ne :: LN : ke = \lambda = \frac{Ne . LN}{ON} = \frac{LN}{ON} y.$$

If, now, we conceive that a force p, acting in the direction of the fibres, or, which is the same thing, acting parallel to the axis of the beam, is applied at k to elongate a single fibre, we have, from equation (3) and the preceding one,

$$p = E \overline{\Delta a} \frac{ke}{LN} = \frac{E}{ON} y \overline{\Delta a},$$

in which $\overline{\Delta a}$ is the transverse section of the fibre. As the section turns about N on the neutral axis, the moment of this force is

$$py = \frac{E}{ON} y^2 \overline{\Delta a}$$

which is found by multiplying the force by the perpendicular y.

This is the moment of a force which is sufficient to elongate or compress any fibre whose original length was LN, an amount equal to the distance between the planes KN and EN measured on the fibre or fibre prolonged. Hence, the sum of all the moments of the resisting forces is

$$\Sigma py = \frac{E}{ON} \Sigma y^2 \overline{\Delta a}$$

in which Σ denotes summation; and in the first member means that the sum of the moments of all the forces which elongate and compress the fibres is to be taken; and in the second member it means that the sum of all the quantities $y^2 \overline{\triangle a}$ included in the transverse section is to be taken. The quantity, $\Sigma y^2 \overline{\triangle a}$ is called the *moment of inertia*, which call I.

But the sum of the moments of the resisting forces equals the sum of the moments of the applied forces. Calling the latter ΣPX, in which X is the arm of the force P, and we have

$$\Sigma py = \Sigma \text{PX} = \frac{\text{E}}{\text{ON}} \Sigma y^2 \overline{\triangle a} = \frac{\text{E.I}}{\text{ON}}$$

$$\therefore \text{ON} = \frac{\text{E.I}}{\Sigma \text{PX}} \quad - \ - \ - \ - \ - \ - \ - \ - \ - \quad (139)$$

In the figure draw Lb tangent to the neutral axis at L, and Na tangent at N. The distance ab, intercepted by those tangents on the vertical through A, is the deflection at A due to the curvature between L and N. As LN is indefinitely short, it may be considered a straight line, and equal x; and Lb = LC very nearly for small deflections; and LC = X. (L stands for two points.)

By the triangles OLN and aLb, considered similar, we have

$$\text{ON} : x :: \text{L}b : ab = \frac{\text{X}x}{\text{ON}}$$

in which substitute ON from equation (139) and we have

$$ab = \frac{\text{X}x\ \Sigma\text{PX}}{\text{E.I}}, \quad - \ - \ - \ - \ - \ - \ - \quad (140)$$

which is sufficiently exact for small deflections. If, now, tangents be drawn at every point of the curve AB, they will divide the line AC into an infinite number of small parts, the sum of which will equal the line AC, the total deflection. But the expression for the value of each of these small spaces will be of the same form as that given above for ab, in which P, E, and I are constant.

This is as far as we can proceed with the general solution. We will now consider

PARTICULAR CASES.

110. CASE 1.—LET THE BEAM BE FIXED AT ONE END, AND A LOAD, P, BE APPLIED AT THE FREE END.—This is a part of Case I., page 99, and Fig. 37 is applicable. The moment of P, in reference to any point on the axis, is PX. Hence ΣPX is simply PX, which, substituted in equation (140), gives

$$ab = \frac{P}{E.I} X^2 x$$

$$\therefore AC = \frac{P}{E.I} \Sigma X^2 x \quad - \; - \; - \; - \; - \; - \; - \; - \; - \quad (141)$$

This equation has been deduced directly from the figure. It now remains to find the sum of all the values of X^2x, which result from giving to X all possible values from $X = 0$ to $X = l$*. To do this, construct a figure some property of which represents the expression, but which has not necessarily any relation to the problem which is being solved. If X be used as a linear quantity, X^2 may be an area and X^2x will be a small volume. These conditions are represented by a pyramid, Fig. 56, in which

$AB = l =$ the altitude, and the base BCDE is a square, whose sides, BC and CD, each $= l$. Let *bcde* be a section parallel to the base, and make another section infinitely near it, and call the distance between the two sections x.

Then $Ab = X = bc = cd$,

$X^2 =$ area *bcde*, and

$X^2x =$ the volume of the lamina *bcde*,

which is the expression sought. The sum of all the laminæ of the pyramid which are parallel to the base is limited by the volume of the pyramid, and this equals the value of the expression $\Sigma X^2 x$ between the limits 0 and l. The volume of the pyramid is the area of the base $(= l^2)$ multiplied by one-third the altitude $(\frac{1}{3}l)$, or $\frac{1}{3}l^3$, which is the value sought.

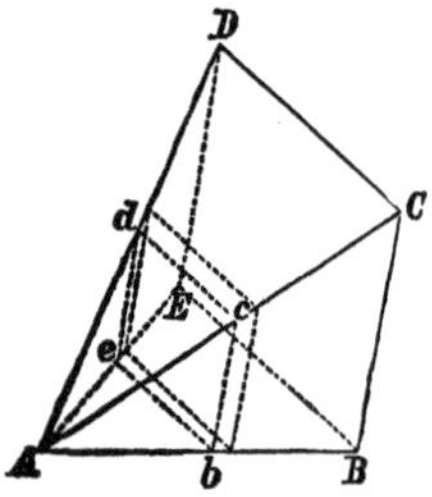

FIG. 56.

Hence, $AC = \dfrac{Pl^3}{3E.I}$

which is the same as equation (57).

The value of X^2x may also be found by statical moments as follows:—Let ABC, Fig. 57, be a triangle, whose thick-

* This by the calculus becomes $\int_0^l x^2\,dx = \frac{1}{3}l^3$

ness is unity, and which is acted upon by gravity (or any other system of parallel forces which is the same on each unit of the body). Take an infinitely thin strip, bc, perpendicular to the base, and let $AB = l = BC$,

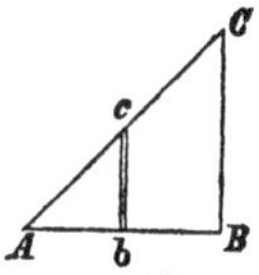

$Ab = X = bc$, and

FIG. 57.

$p =$ the weight of a unit of volume.

Then $Xx =$ the area of the infinitely thin strip bc, and

$pXx =$ the weight of the strip bc, and

$pX^2x =$ the moment of the strip, when A is taken as the origin of moments. If the weight of a unit of volume be taken as a unit, the moment becomes X^2x, which is the quantity sought, and the value of ΣX^2x from 0 to l is the moment of the whole triangle ABC. Its area is $\frac{1}{2}l^2$, and its centre of gravity $\frac{2}{3}l$ to the right of A. Hence the moment is $\frac{1}{3}l^3$ as before found.*

111. CASE II.—LET THE BEAM BE FIXED AT ONE END, AND UNIFORMLY LOADED OVER ITS WHOLE LENGTH.— This is the same as a part of Case II., page 101, and Fig. 39 is applicable.

Let X be measured from the free end, and

$w =$ the load on a unit of length; then

$wX =$ the load on a length X, and

$\frac{1}{2}X =$ the distance of the centre of gravity of the load from the section which is considered.

Hence the moment is $\frac{1}{2}wX^2$, which equals ΣPX, and equation (140) becomes

$$ab = \frac{w}{2E.I} X^3x, \text{ and}$$

$$AC = \frac{w}{2E.I} \sum_{x=0}^{x=l} X^3x = \text{the total deflection.}$$

To find the value of ΣX^3x, observe, in Fig 56, that X^2x is the volume of the lamina $bcde$, and this multiplied by the altitude of $A - bcde$, which is X, gives X^3x, the expression sought. Hence the sum sought is the *volume* of the pyramid $A - BCDE$,

---

* This may be written $\sum_{x=o}^{x=l} X^2x = \frac{1}{3}l^3$.

multiplied by the distance of the centre of gravity of the pyramid from the apex; or,

$$l^2 \times \tfrac{1}{3}l \times \tfrac{3}{4}l = \tfrac{1}{4}l^4$$

$$\therefore \text{AC} = \frac{wl^4}{8\text{E.I}} = \frac{\text{W}l^3}{8\text{E.I}} \quad - \; - \; - \; - \quad (142)$$

where W is the total load on the beam.

112. CASE III.—LET THE BEAM BE SUPPORTED AT ITS ENDS AND LOADED AT THE MIDDLE BY A WEIGHT P, as in Fig. 40. The reaction of each support is $\frac{1}{2}$P, and the moment is $\frac{1}{2}$PX, and equation (140) becomes

$$ab = \frac{\text{P}}{2\text{E.I}}\text{X}^2x.$$

But in this case the greatest deflection is at the middle, and the limits of ΣX^2x are 0 and $\frac{1}{2}l$. Hence, in Fig. 56, let the altitude of the pyramid be $\frac{1}{2}l$, and each side of the base also $\frac{1}{2}l$, and the volume will be

$$\tfrac{1}{2}l \times \tfrac{1}{2}l \times \tfrac{1}{3} \text{ of } \tfrac{1}{2}l = \tfrac{1}{24}l^3$$

$$\therefore \text{AC} = \frac{\text{P}l^3}{48\text{E.I}},$$

which is the same as equation (73).

113. CASE IV.—LET THE BEAM BE SUPPORTED AT ITS ENDS AND UNIFORMLY LOADED, AS IN FIG. 42.

w being the load on a unit of length, the reaction of each support is $\frac{1}{2}wl$, and its moment at any point of the beam is $\frac{1}{2}wl\text{X}$. On the length X there is a load wX, the centre of which is at $\frac{1}{2}$X from the point considered; hence its moment is $\frac{1}{2}w\text{X}^2$, and the total moment is the difference of these moments;

$$\therefore \Sigma \text{PX} = \tfrac{1}{2}wl\text{X} - \tfrac{1}{2}w\text{X}^2,$$

and equation (140) becomes

$$ab = \frac{w}{2\text{E.I}}\,(l\text{X}^2x - \text{X}^3x),$$

and the total deflection at the middle is,

$$\text{AC} = \frac{w}{2\text{E.I}}\Big(l\sum_{x=0}^{x=\frac{1}{2}l}\text{X}^2x - \sum_{x=0}^{x=\frac{1}{2}l}\text{X}^3x \Big).$$

The values of the terms within the parentheses have already been found, and by subtracting them we have

$$AC = \frac{5}{384}\frac{Wl^3}{E.I}.$$

114. REMARK ABOUT OTHER CASES.—This method, which appears so simple in these cases, unfortunately becomes very complex in many other cases, and in some it is quite powerless. To solve the 9th and 10th cases, pages 115 and 116, necessitates an expression for the inclination of the curve, so that the condition of its being horizontal over the support may be imposed upon the analysis. But the 9th case may be easily solved if we find by any process that the weight which must be suspended at the outer end of the beam to make it horizontal over the support is $\frac{1}{8}Pl$ divided by AC,* Fig. 47. For, the reaction of the support is $\frac{1}{2}P + P$;

$$\begin{aligned} \therefore \Sigma PX &= P_1(AC + X) - (\tfrac{1}{2}P + P_1)X \\ &= P_1 AC - \tfrac{1}{2}PX \\ &= \tfrac{1}{8}Pl - \tfrac{1}{2}PX \end{aligned}$$

$$\therefore ab = \frac{1}{8}\frac{PlXx - 4PX^2x}{E.I},$$

and the deflection at the centre $= \frac{1}{8}\frac{P}{E.I}(l\Sigma Xx - 4\Sigma X^2x)$ taken between the limits 0 and $\frac{1}{2}l$.

The part ΣXx is the area of a triangle whose base and altitude are each $\frac{1}{2}l$, $\therefore \Sigma Xx = \frac{1}{8}l^2$, and ΣX^2x between the limits 0 and $\frac{1}{2}l$, is $\frac{1}{24}l^3$ $\therefore$ AC† $= \frac{P}{192}\frac{l^3}{E.I}$.

All these expressions contain I, the value of which remains to be found by the graphical method.

---

* This "AC" refers to Fig. 47.

† This "AC" refers to Fig. 55.

115. MOMENT OF INERTIA OF A RECTANGLE.—*Required the moment of inertia of a rectangle about one end as an axis.*

Let ABCD, Fig. 58, be a rectangle. Make BG perpendicular to and equal AB, and complete the wedge G – ABCD.

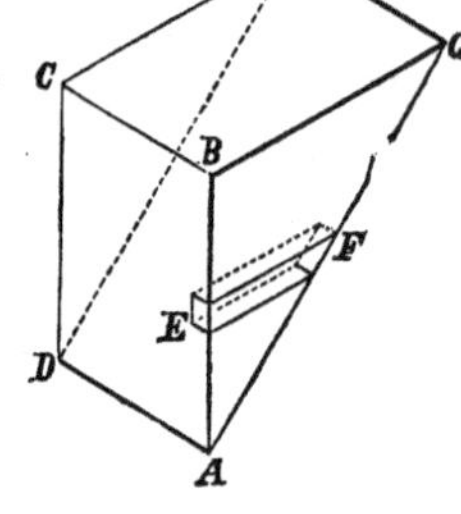

FIG. 58.

Let $\overline{\triangle a}$ = the area of a very small surface at E, and

y = AE = EF, then

$y\overline{\triangle a}$ = the volume of a very small prism EF, and this multiplied by y gives

$y^2\overline{\triangle a}$ = the moment of inertia of the elementary area at E, which is also the statical moment of the prism EF, and

$\Sigma y^2\overline{\triangle a}$ = I = the moment of inertia of the rectangle ABCD.

Hence the moment of inertia of the rectangle is *represented* by the statical moment of the wedge G – ABCD. If

$$\text{AB} = d = \text{BG, and}$$
$$\text{AD} = b,$$

then the volume of the wedge is

$$bd \times \tfrac{1}{2}d = \tfrac{1}{2}bd^2$$

and the moment $= \frac{1}{2}bd^2 \times \frac{2}{3}d = \frac{1}{3}bd^3$ - - - - - - (143)

If the axis of moments passes through the centre of the rectangle, and parallel to one end, we have BE = GB = $\frac{1}{2}d$ in Fig. 59. Hence the moment of inertia of the rectangle =

$$2 \times b \times \tfrac{1}{2}d \times \tfrac{1}{4}d \times \tfrac{2}{3} \text{ of } \tfrac{1}{2}d = \tfrac{1}{12}\,bd^3$$

which is the same as equation (50).

116. THE MOMENT OF INERTIA OF A TRIANGLE about an axis parallel to the base and passing through the vertex is, in a similar way, the statical moment of the pyramid ABCDE. Fig. 60.

Let b = CB = base of the triangle, and

d = AB = BD = CE = altitude of the triangle and pyramid and sides of the base of the pyramid.

The volume of the pyramid $= \frac{1}{3}\,bd^2$.

The centre of gravity is $\frac{3}{4}d$ from the apex, consequently the statical moment is $\frac{1}{3}\,bd^2 \times \frac{3}{4}d = \frac{1}{4}bd^3$.

But in a triangular beam the neutral axis passes through the centre of gravity of the triangle, and it is desirable to find the moment of inertia about an axis which passes through the centre and parallel to the base.

This may be done as in the preceding Article; but it may be

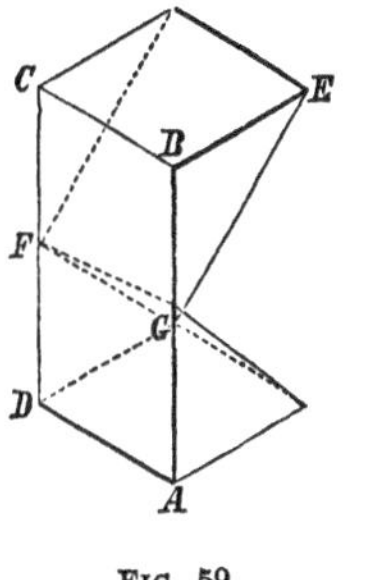

FIG. 59.

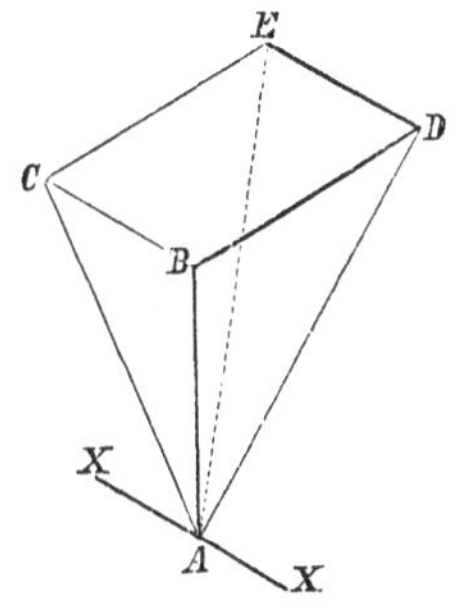

FIG. 60.

more easily done by using the *formula of reduction*, which is as follows:—*The moment of inertia of a figure about an axis passing through its centre equals the moment of inertia about an axis parallel to it, minus the area of the figure multiplied by the square of the distance between the axes.* (See Appendix III.)

This gives for the moment of inertia of a triangle about an axis passing through its centre and parallel to the base

$$\tfrac{1}{4}bd^3 - \tfrac{1}{2}bd \times (\tfrac{2}{3}d)^2 = \tfrac{1}{36}\,bd^3 \quad - \quad - \quad - \quad - \quad - \quad - \quad (143)$$

117. THE MOMENT OF INERTIA OF A CIRCLE may be represented in the same way, but it is not easy to find the volume of the wedge, or the position of its centre of gravity, except by analysis which is more tedious than that required to find the moment directly, as was done in equation (51). But it may be found practically, by those who can only perform multiplication, as follows:—Make a wedge-shaped piece out of wood, or plaster of Paris, or other convenient material, the base of which is the semicircle required, and the altitude is the radius of the circle; then find its volume by immersing it in a liquid and measuring the amount of water displaced. Then determine the distance of the centre of gravity of the wedge from the centre of the circle by balancing it on a knife edge, holding the edge of the knife under the base of the wedge, and parallel to the edge, *ab*, of the wedge, keeping the side

FIG. 60 *a*.

vertical, and measuring the distance between the edge ab and the line of support. Then the statical moment is the product of the volume multiplied by the horizontal distance of the centre from the edge. Its value for the whole circle, or for both wedges, is $\frac{1}{4}\pi r^4$.

There are, however, many methods of calculating the moment of inertia of a circle without using the Calculus. The following method, which the author has devised, appears as simple as any of the known methods :—

The moment of inertia of a circle is the same about all its diameters. Hence the moment about X in the figure, *plus* the moment about Y, equals *twice* the moment about X. The distance to any point A is ρ, and equals $\sqrt{x^2 + y^2}$; or $\rho^2 = x^2 + y^2$; and if $\overline{\triangle a}$ be an elementary area, as before, we have

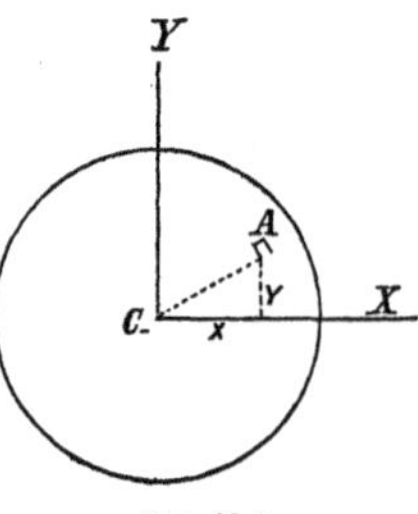

FIG. 60 b.

$$2\Sigma\overline{\triangle a}\, x^2 = \Sigma\overline{\triangle a}\, x^2 + \Sigma\overline{\triangle a} y^2 = \Sigma\overline{\triangle a}\, \rho^2,$$

the latter of which is called the polar moment of inertia, in reference to an axis perpendicular to the plane of the circle, and passing through its centre C. To find the value of $\Sigma\overline{\triangle a}\, \rho^2$, take a triangle whose base and altitude are each equal to r, the radius of the circle, and revolve it about the axis through C, and construct an infinitely small prism on the element $\overline{\triangle a}$ as a base.

We have ρ = CA = AB, Fig. 60 *c*.

$\overline{\triangle a}\, \rho$ = volume of the small prism *AB*.

$\overline{\triangle a}\, \rho\,$CA $= \overline{\triangle a}\, \rho^2$ = the moment of *AB*, the form of the quantity sought.

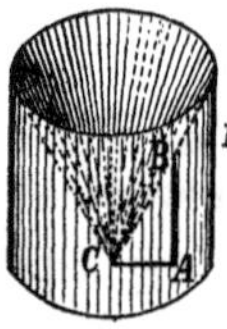

FIG. 60 c.

Hence $\Sigma\overline{\triangle a}\, \rho^2$ is the product of the volume of the solid generated by the triangle, multiplied by the abscissa of its centre of gravity from C. The solid is what remains of a cylinder after a cone has been taken out of it, the base of the cone being the upper base of the cylinder, and the apex of which is at the centre of the base of the cylinder. Hence the volume of the solid is the volume of the cylinder, *less* the volume of the cone; or $\pi r^2 \times r - \pi r^2 \times \frac{1}{3}r = \frac{2}{3}\pi r^3$.

If now the solid be divided into an infinite number of pieces, by planes which pass through its axis, each small solid will be a pyramid, having its vertex at C, and the abscissa to the centre of gravity of each is $\frac{3}{4}r$ from C. Hence we finally have

$$\Sigma \overline{\triangle a}\, \rho^2 = \tfrac{2}{3}\pi r^3 \times \tfrac{3}{4} r = \tfrac{1}{2}\pi r^4, \text{ which}$$

equals $2\, \Sigma \overline{\triangle a}\, x^2$.

$$\therefore \Sigma \overline{\triangle a}\, x^2 = \tfrac{1}{4}\pi r^4. \quad - \; - \; - \; - \; - \; - \; - \quad (144)$$

118. MOMENT OF INERTIA OF OTHER SURFACES.—The general method indicated in the preceding articles is applicable to surfaces of any character, and with careful manipulation approximations may be made which will be very nearly correct, and, as we have seen above, in some cases exact formulas may be found.

CHAPTER VI.

TRANSVERSE STRENGTH.

119. STRENGTH OF RECTANGULAR BEAMS.—The theories which have been advanced from time to time to explain the mechanical action of the fibres, have been already given in Chapter IV. Both the common theory, and Barlow's theory of "the resistance to flexure" will be considered in this chapter.

First, consider the common theory, according to which the neutral axis passes through the centre of gravity of the transverse sections, and the strain upon the fibres is directly proportional to their distance from the neutral axis.

Continuing the use of the geometrical method, let Fig. 61 represent a rectangular beam which is strained by a force P applied at any point. Let *de* be on the neutral axis, and *ab* represent the strain upon the lowest fibre. Pass a plane, *de–cb*, and the wedge so cut off *represents* the strains on the lower side, and the similar wedge on the other side *represents* the strains on the upper side.

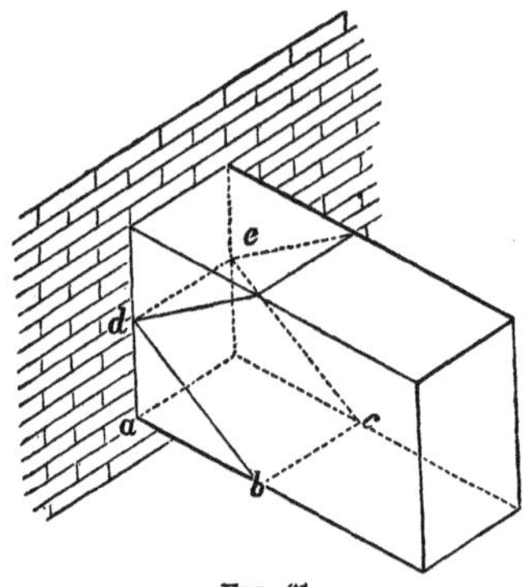

FIG. 61.

Let R = the strain upon a unit of fibres most remote from the neutral axis on the side which first ruptures, on the hypothesis that all the fibres of the unit are equally strained, and b = the breadth and d = the depth of the beam.

Let $ab = \mathrm{R}$; then, the total resistance to compression $= \frac{1}{2}\mathrm{R}b \times \frac{1}{2}d = \frac{1}{4}\mathrm{R}bd$, = the volume of the lower wedge; and the *mo-*

ment of resistance is this value multiplied by the distance of the centre of gravity of the wedge from *de*, which is $\frac{2}{3}$ of $\frac{1}{2}d = \frac{1}{3}d$; consequently the *moment* is

$$\tfrac{1}{12}Rbd^2;$$

and as the moment of resistance to tension is the same, *the total moment of resistance is*

$$\tfrac{1}{6}Rbd^2, \quad - \; - \; - \; - \; - \; - \; - \; - \; - \; - \; - \quad (145)$$

which equals the moment of the applied or bending forces.

If the beam be fixed at one end and loaded by a weight, P, at the free end, we have for the *dangerous section*, or that most liable to break,

$$Pl = \tfrac{1}{6}Rbd^2.$$

In rectangular beams the dangerous section will be where the sum of the moments of stresses is greatest, the maximum values of which for a few cases are given in a table on page 118. Using those values, and placing them equal to $\frac{1}{6}Rbd^2$, and we have for solid rectangular beams at the dangerous section, the following formulas:—

FOR A BEAM FIXED AT ONE END AND A LOAD, P, AT THE FREE END; $Pl = \frac{1}{6}Rbd^2$; - - - - - - - - - - - - (146)

AND FOR AN UNIFORM LOAD; $\frac{1}{2}Wl = \frac{1}{6}Rbd^2$ - - (147)

FOR A BEAM SUPPORTED AT ITS ENDS AND A LOAD, P, AT THE MIDDLE; $\frac{1}{4}Pl = \frac{1}{6}Rbd^2$; - - - - - - - - - - (148)

AND FOR AN UNIFORM LOAD; $\frac{1}{8}Wl. = \frac{1}{6}Rbd^2$; - (149)

AND FOR A LOAD AT THE MIDDLE, AND ALSO AN UNIFORM LOAD; $\frac{1}{8}(2P + W)l = \frac{1}{6}Rbd^2$ - - - - - - - - - - - - (150)

FOR A BEAM FIXED AT BOTH ENDS AND A LOAD, P, AT THE MIDDLE; $\frac{1}{8}Pl = \frac{1}{6}Rbd^2$; - - - - - - - - - - (151)

AND FOR AN UNIFORM LOAD, END SECTION; $\frac{1}{12}Wl = \frac{1}{6}Rbd^2$; (152)

MIDDLE SECTION; $\frac{1}{24}Wl = \frac{1}{6}Rbd^2$ - - - - - - (153)

These expressions show that in solid rectangular beams the strength varies as the breadth and square of the depth, and hence breadth should be sacrificed for depth. In all the cases, except for a beam fixed at the ends, it appears that a beam will support twice as much if the load be uniformly distributed over the whole length as if it be concentrated at the middle of the length. The case in which a beam is fixed at both ends and loaded at the middle has given rise to considerable discussion, for it is found by experiment that a beam whose ends are fixed in walls or masonry will not sustain as much as is indicated by the formula, and also that it requires considerably more load to break it at the ends than at the middle, but the analysis shows that it is equally liable to break at the ends or at the middle. But it should be observed that there is considerable difference between the condition of mathematical fixedness, in which case the beam is horizontal over the supports, and that of imbedding a beam in a wall. For in the latter case the deflection will extend some distance into the wall.

Mr. Barlow concludes from his experiments that equation (151) should be

$$\tfrac{1}{6}Pl = \tfrac{1}{6}Rbd^2 \quad \text{- - - - - - - - -} \quad (154)$$

and this relation is doubtless more nearly realized in practice than the ideal one given above. In either case, it appears that writers and experimenters have entirely overlooked the effect due to the change of position of the neutral axis, which must take place. It has been assumed that the neutral axis coincides with the axis of the beam, and that its length remains unchanged during flexure; but if the ends of the beam are fixed, the axis must be elongated by flexure, or else approach much nearer the concave than the convex side, or both take place at the same time, in which case the *moment of resistance* will not be $\tfrac{1}{6}Rbd^2$. The phenomena are of too complex a character to admit of a thorough and exact analysis, and it is probably safer to accept the results of Mr. Barlow in practice than depend upon theoretical results.

120. MODULUS OF RUPTURE.—When a beam is supported at its ends, and loaded uniformly over its whole length, and also loaded at the middle, we find from equation (150)

$$R = \frac{\tfrac{3}{4}(2P + W)}{bd^2}l \quad \text{- - - - - - - -} \quad (155)$$

in which W may be the weight of the beam. Beams of known dimensions, thus supported, have been broken by weights placed at the middle of the length, and the corresponding value of R has been found for various materials, the results of which have been entered in the table in Appendix IV. This is called the MODULUS OF RUPTURE, and is defined to be *the strain upon a square inch of fibres most remote from the neutral axis on the side which first ruptures.* It would seem from this definition that R should equal either the tenacity or crushing resistance of the material, depending upon whether it broke by crushing or tearing, but an examination of the table shows the paradoxical result that it never equals either, but is always greater than the smaller and less than the greater. For instance, in the case of cast iron:—

The mean value of T = 16,000 lbs.
" " C = 96,000 "
" " R = 36,000 " nearly,

hence R is about $2\frac{1}{4}$ times T, and a little over $\frac{1}{3}$ of C.

For English oak—

T = 17,000 lbs.;
C = 9,500 lbs.; and
R = 10,000 lbs.;

hence R exceeds C, and is more than half of T.

For ash—

T = 17,000 lbs.;
C = 9,000 lbs.; and
R = 12,000 lbs.; hence
R = $1\frac{1}{3}$ C and about $\frac{2}{3}$ T.

These discrepancies have long been recognized, and the cause has generally been attributed to a departure from the law of perfect elasticity and a movement of the neutral axis away from the centre of the beam in the state bordering on rupture; but as the laws of these variations were not assigned, their influence could not be analyzed. (See Articles 74 and 75.)

The tabulated values of R being found from experiments upon solid rectangular beams, they are especially applicable to all beams of that form, and they answer for all others that do not depart largely from that form; but if they depart largely from that form, as in the case of the I (double T) section, or hollow beams, or other irregular forms, the formulas will give results somewhat in excess of the true strength; and in such cases Barlow's theory gives results more nearly correct.

But if, instead of R, we use T or C, whichever is smaller, in the formulas which we have deduced, and suppose that the neutral axis remains at the centre of the beam, *we shall always be on the safe side;* but there would often be an excess of strength, as, for instance, in the case of cast iron the actual strength of the beam would be about twice as strong as that found by such a computation.

The difficulty is avoided, practically, by using such a small fractional part of R as that it will be considered perfectly safe. This fraction is called the *coefficient of safety.* The values commonly used for beams are the same as for bars, and are given in Article 38.

Experiments should be made upon the material to be used in a structure, in order to determine its strength; but in the absence of such experiments the following mean values of R are used:—

850 to 1,200 lbs. for wood,
10,000 to 15,000 lbs. for wrought iron, and
6,000 to 8,000 lbs. for cast iron.

121. PRACTICAL FORMULAS.

If R = 1,000 for wood, and
12,000 for wrought iron,

we have for a rectangular beam, supported at its ends and loaded at the middle of its length,

$$P = \frac{666\ bd^2}{l} \text{ for wooden beams; and}$$

$$P = \frac{8000\ bd^2}{l} \text{ for wrought-iron beams.}$$

The length of the beam, and the load it is to sustain, are generally known quantities, and the breadth and depth are required; but it is also necessary to assume one of the latter, or assign a relation between them. For instance, if the depth be n times the breadth, the preceding formulas give

$$b = \sqrt[3]{\frac{Pl}{666n^2}}\ ;\ \text{and } d = \sqrt[3]{\frac{Pln}{666}} \text{ for wood;} \qquad (156)$$

$$\text{and } b = \sqrt[3]{\frac{Pl}{8000n^2}}\ ;\ \text{and } d = \sqrt[3]{\frac{Pln}{8000}} \text{ for wrought iron;} \qquad (157)$$

122. THE RELATIVE STRENGTH OF A BEAM under the various conditions that it is held is as the moment of the applied forces; hence, all the cases which have been considered may, *relatively*, be reduced to *one*, by finding how much a beam will carry which is fixed at one end and loaded at the free end, equation (146), and multiplying the results by the following factors:—

	FACTORS.
Beam fixed at one end and loaded at the other - - -	1
" " " " uniformly loaded - - - -	2
Beam supported at its ends and loaded at the middle -	4
" " " " uniformly loaded - - -	8
Beam fixed at one end and supported at the other, and uniformly loaded - - - - - - - - - - - -	8
Beam fixed at both ends and loaded at the middle - -	8
" " " " uniformly loaded - - -	12

If it is required to know the breadth of a beam which will sustain a given load, find b, from equation (146), and for a beam in any other condition, divide by the factors given above for the corresponding case.

If the depth is required, find d from equation (146), and divide the result for the particular case desired by the square root of the above factors.

123. EXAMPLES.—

1. A beam, whose depth is 8 inches, and length 8 feet, is supported at its ends, and required to sustain 500 pounds per foot of its length; required its breadth so that it will have a factor of safety of $\frac{1}{10}$, R being 14,000 pounds.

From equation (146) we have,

$$b = \frac{6Pl}{Rd^2} = \frac{6 \times 500 \times 8 \times 8 \times 12}{1400 \times 8^2} = 25\tfrac{5}{7} \text{ inches};$$

and by examining the above table of factors we see that this must be divided by 8; $\therefore$ Ans. $= 3\frac{3}{14}$ inches.

2. If $l = 10$ feet, P at the middle $= 2,000$ lbs., $b = 4$ inches, $R = 1,000$ lbs., required d.

3. If a beam, whose length is 8 feet, breadth is 3 inches, and depth 6 inches,

is supported at its ends, and is broken by a weight of 10,000 pounds placed at the middle, and the weight of a cubic foot of the beam is 50 pounds; required the value of R. Use equation (150).

4. If $R = 80,000$ lbs., $l = 12$ feet, $b = 2$ inches, $d = 5$ inches, how much will the beam sustain if supported at its ends and loaded uniformly over its whole length, coefficient of safety $\frac{1}{4}$? Ans $W = 9,259$ lbs.

5. A wooden beam, whose length is 12 feet, is supported at its ends; required its breadth and depth so that it shall sustain one ton, uniformly distributed over its whole length. Let $R = 15,000$ lbs., coefficient of safety $\frac{1}{10}$, and depth = 4 times the breadth. Ans. $b = 2.08$ inches.
$d = 8.32$ inches.

6. A beam is 2 inches wide and 8 inches deep, how much more will it sustain with its broad side vertical, than with it horizontal?

7. A wrought-iron beam 12 feet long, 2 inches wide, 4 inches deep, is supported at its ends. The material weighs $\frac{1}{4}$ lb. per cubic inch; how much load will it sustain uniformly distributed over its whole length, $R = 54,000$ lbs.? Ans. Without the weight of the beam, 15,712 lbs.

8. A beam is fixed at one end; $l = 20$ feet, $b = 1\frac{1}{2}$ inch, $R = 40,000$ lbs.; weight of a cubic inch of the beam $\frac{1}{4}$ lb. Required the depth that it may sustain its own weight and 500 lbs. at the free end. Ans. 4.05 inches.

9. The breadth of a beam is 3 inches, depth 8 inches, weight of a cubic foot of the beam 50 pounds, $R = 12,000$; required the length so that the beam shall break from its own weight when supported at its ends.
Ans. $l = 175.27$ feet.

124. RELATION BETWEEN STRAIN AND DEFLECTION.—When the strain is within the elastic limit we may easily find the greatest strain on the fibres corresponding to a given deflection. For instance, take a rectangular beam, supported at its ends and loaded at the middle of its length, and we have from equation (148)

$$P = \frac{2}{3}\frac{Rbd^2}{l}$$

and from equations (73) and (51)

$\Delta = \frac{1}{4}\frac{Pl^3}{Ebd^3}$, which becomes, by substituting P from the preceding,

$$\Delta = \frac{1}{6}\frac{Rl^2}{Ed}$$

$$\therefore R = \frac{6Ed}{l^2} \Delta \quad - - - - - - - - - - - - \quad (158)$$

Examples.—1. If $l = 6$ feet, $b = 1\frac{1}{2}$ inch, $d = 4$ inches, coefficient of elasticity $= 25,000,000$ lbs. is supported at its ends and loaded at the middle so as to produce a deflection at the middle of $\Delta = \frac{3}{4}$ inch; required the greatest strain on the fibres. Also required the load.

2. On the same beam, if the greatest strain is $R = 12,000$ lbs., required the greatest deflection.

3. If the beam is uniformly loaded, required the relation between the greatest strain and the greatest deflection.

4. Generally, prove that $R = \text{constant} \times \frac{Ed}{l^2} \Delta$.

125. HOLLOW RECTANGULAR BEAMS.—If a rectangular beam has a rectangular hollow, both symmetrically placed in reference to the neutral axis, as in Fig. 62, we may find its strength by deducting from the strength of a solid rectangular beam the strength of a solid beam of the same size as the hollow. But in this case, when the beam ruptures at b, the strain at b' will be less than R. As the strains increase directly as the distance of the fibres from the neutral axis, we have, if d and d' are the depths of the outside and hollow parts respectively,

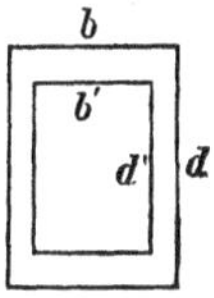

FIG. 62.

$$\tfrac{1}{2}d : \tfrac{1}{2}d' :: R : \text{strain at } b' = R\frac{d'}{d}.$$

If b' = the breadth of the hollow, the stress on that part, if it were solid, would be, according to equation (145),

$$\tfrac{1}{6}\left(R\frac{d'}{d}\right) b'd'^2 = \tfrac{1}{6}R\frac{b'd'^3}{d},$$

which, taken from equation (145), gives for the resistance of a hollow rectangular beam,

$$\tfrac{1}{6}R\frac{bd^3 - b'd'^3}{d} \quad - - - - - - - - - - \quad (159)$$

If the hollow be on the outside, as in Fig. 63, forming an H section, the result is the same.

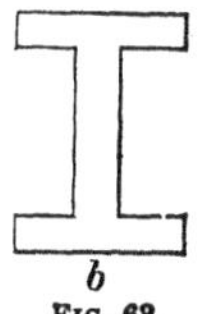

FIG. 63.

126. IF THE UPPER AND LOWER FLANGES ARE UNEQUAL it forms a double T, as in Fig. 64. Let the notation be as in the figure, and also d_1 equal the distance from the neutral axis to the upper element, and x the distance from the neutral axis to the lower element.

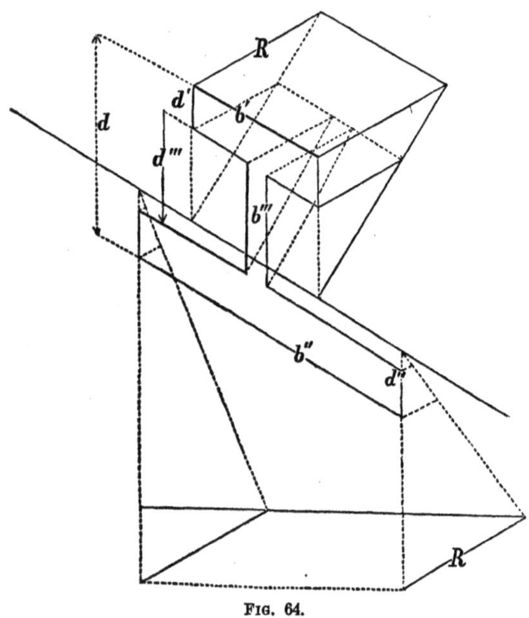

FIG. 64.

To find the position of the neutral axis, make the statical moments of the surface above it equal to those below it. This gives

$$d' b' (d_1 - \tfrac{1}{2} d') + \tfrac{1}{2} b''' (d_1 - d')^2 = d'' b'' (x - \tfrac{1}{2} d'') + \tfrac{1}{2} b''' (x - d'')^2 \qquad (160)$$

We also have $d_1 = d - x = d' + d'' + d''' - x$ - (161)

These equations will give x and d_1.

Constructing the wedges as before, and the resistance to compression is represented by the wedge whose base is $b' d_1$ and altitude R, *minus* the wedge whose base is $(b' - b''') (d_1 - d')$ and altitude $\frac{d_1 - d'}{d_1}$ R. Hence the resistance to compression is

$$\tfrac{1}{2} R b' d_1 - \tfrac{1}{2} \frac{d_1 - d'}{d_1} R (b' - b''') (d_1 - d')$$

The centre of gravity is at $\frac{3}{8}$ the altitude, or $\frac{3}{8}d_1$ for the former wedge, and $\frac{3}{8}(d_1 - d')$ for the latter, and if the volumes be multiplied by these quantities respectively, it will give for *the moment of resistance to compression*

$$\tfrac{1}{8}Rb'd_1^2 - \tfrac{1}{8}\frac{R}{d_1}(b' - b''')(d_1 - d')^3$$

Next consider the resistance to tension. Since the strains on the elements are proportional to their distances from the neutral axis, therefore

$$d_1 : x :: R : \text{strain at the lower side of the section} = \frac{R}{d_1}x,$$

and similarly,

$d_1 : (x - d'') :: R :$ strain at the opposite side of the lower flange $= \dfrac{R}{d_1}(x - d'')$.

Hence the tensive strains will be represented by a wedge whose base is $b''x$ and altitude $\dfrac{R}{d_1}x$, *minus* a wedge whose base is $b'' - b'''$ and altitude $\dfrac{R}{d_1}(x - d'')$. Hence the moment of resistance is

$$\tfrac{1}{8}\frac{R}{d_1}b''x^3 - \tfrac{1}{8}\frac{R}{d_1}(b'' - b''')(x - d'')^3$$

The total moment of resistance is the sum of the two moments, or

$$\tfrac{1}{8}\frac{R}{d_1}\Big[b'd_1^3 - (b' - b''')(d_1 - d')^3 + b''x^3 - (b'' - b''')(x - d'')^3\Big] \quad \text{- - - - - - - - - -} \quad (162)$$

For a single T make b'' and $d'' = 0$ in the above expression.

The method which has here been applied to rectangular beams may be applied to beams of any form; but it often requires a knowledge of higher mathematics to find the volume of the wedge, and the position of its centre of gravity; or resort must be had to ingenious methods in connection with actual wedges of similar dimensions.

127. TRUE VALUE OF d_1 AND AN EXAMPLE.—In this and similar expressions

d_1 = *the distance from the neutral axis to the fibre most remote from it* ON THE SIDE WHICH FIRST RUPTURES.

d_1 is usually taken as the distance to the most remote fibre, without considering whether rupture will take place on that side or not; but this oversight may lead to large errors.

For example, let the dimensions of a cast-iron double T beam be as in Fig. 65, and 228 inches between the supports. Required the load at the middle necessary to break it.

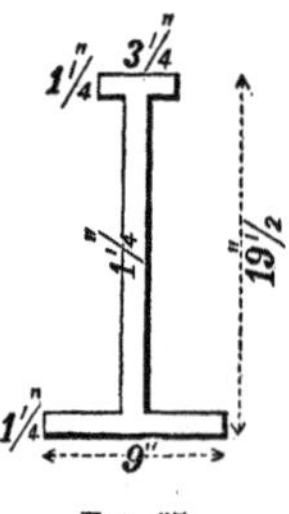

FIG. 65.

The position of the neutral axis is found from equations (160) and (161) to be 7.96 inches from the lower side, and 11.54 inches from the upper. As cast iron will resist from four to six times as much to compression as to tension—this beam will rupture on the lower side first; hence d_1 in the equation = 7.96 inches. As the value of R is not known, take a mean value = 36,000 lbs. The moment of the rupturing force—neglecting the weight of the beam—is $\frac{1}{4}$ Pl, which placed equal to expression (162) and reduced gives

$$P = \frac{4}{228} \times \frac{36,000}{7.96} \times 1,672 = 132,000 \text{ lbs.} = 58.9 \text{ tons gross.}$$

Had we used d_1 = 11.54, it would have given P = 40.4 tons. Such beams actually broke with from 50 to 54 tons; or, including the weight of the beam, with a mean value of 52½ tons.

By reversing the problem, and using 52½ tons for P, we find that R is a little more than 32,000 pounds. Had this value of R been used in the first solution, and d_1 made equal 11.54, it would have given for P a little more than 36 tons, which would be the strength if the beam were inverted. If the upper flange were smaller or the lower larger, the discrepancy would have been greater.

The strain upon a fibre in the upper surface is to the strain upon one in the lower surface as d_1 to x; hence, if the material resists more to compression than to tension (as cast iron), it should be so placed that the small flange shall resist the former, and the large one the latter. If a cast-iron beam be supported at its ends, the smaller flange should be uppermost, and as it resists from four to six times as much compression as tension, the neutral axis should be from four to six times as far from the upper surface as from the lower, for economy. Using the same notation as in Fig. 64 and we have,

$$\frac{d_1}{x} = \frac{\text{greatest compressive strain}}{\text{greatest tensive strain}},$$

and for economy we should have,

$$\frac{d_1}{x} = \frac{\text{ultimate compressive strength}}{\text{ultimate tensile strength}}.$$

The ultimate resistance of wrought iron is greater for tension than for compression; hence, if a wrought-iron beam is supported at its ends, the heavier flange should be uppermost.

The proper thickness of the vertical web can be determined only by experiment, and this has been done, in a measure, by Baron von Weber, in his experiments on permanent way.

128. EXPERIMENTS OF BARON VON WEBER for determining the thickness required for the central web of rails.

Baron von Weber desired to ascertain what was the *minimum* thickness which could be given to the web of a rail, in order that the latter might still possess a greater power of resistance to lateral forces than the fastenings by which it was

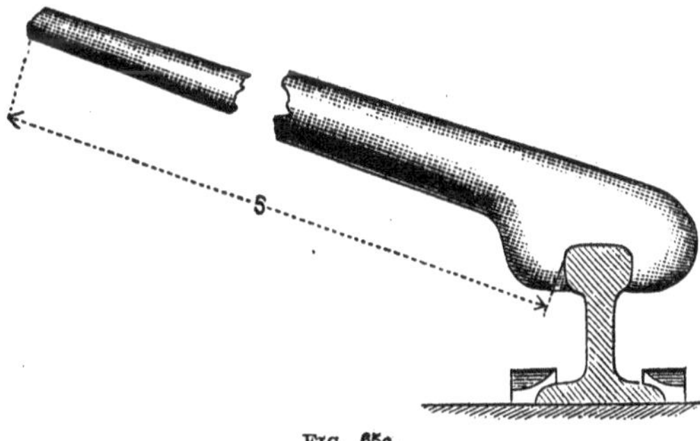

FIG. 65a.

secured to the sleepers. For this purpose a piece of rail 6 feet in length, rolled, of the best iron at the Laurahutte, in Silesia, was supported at distances of 35.43 in., and loaded nearly to the limit of elasticity (which had been determined previously by experiments on other pieces of the same rail), and the deflections were then measured with great care by an instrument capable of registering 1-1000 in. with accuracy. This having been done, the web of the piece of rail was planed down, and each time that the thickness had been reduced 3 millimetres the vertical deflection of the rail under the above load was again tested, and the rail was subjected to the following rough but practical experiments. The piece of rail was fastened to twice as many fir sleepers by double the number of spikes which would be employed in practice, and a lateral pressure was then applied to the head of the rail by means of a lifting-jack, until the rail began to cant and the spikes were drawn. The same thing was then done by a sudden pull, the apparatus used being a long lever fastened to the top of the rail, as shown in Fig. 65a. The lifting-jack and the lever were applied to the ends of the rail, and the

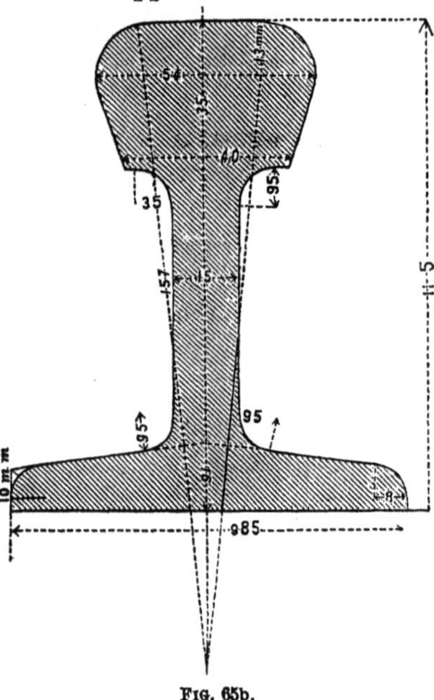

FIG. 65b.

web of the latter had, in each case, to resist the whole strain required for drawing out the spikes. The results of the experiments made to ascertain the resistance of the rail to vertical flexure with different thicknesses of web, and under a load of 5,000 lbs., were as follows:—

	Thickness of web. In.	Vertical deflection. In.
15 millimetres	= 0.59	0.016
12 "	0.47	0.016
9 "	0.35	0.019
6 "	0.24	0.0194
3 "	0.12	0.022

These results showed ample stiffness, even when the web was reduced in thickness to 0.12 in. To determine the power of resistance of the rail to lateral flexure, an impression of the section was taken in lead each time that the spikes were drawn.

The forces applied in these experiments were very far greater than those occurring in practice, yet it was found that with the web 12, 9, and even 6 millimetres thick, no distortion took place, and only when the thickness of the web was reduced to 3 millimetres (0.12 in.) was a slight permanent lateral deflection of the head caused just as the spikes gave way. The section shown in Fig. 65b had then been reduced to that shown in Fig. 65c.

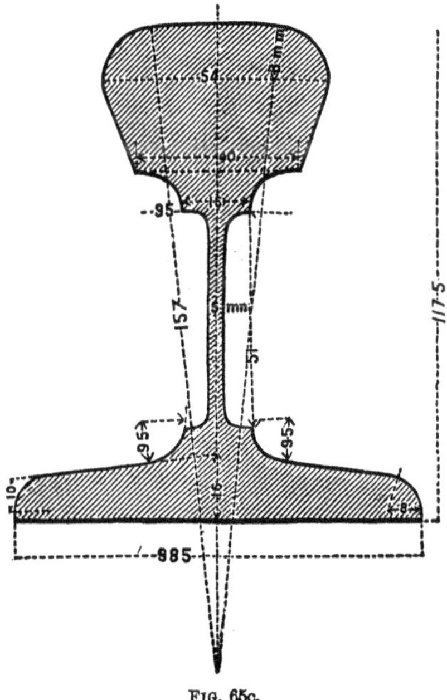

Fig. 65c.

Next, a rail, with the web reduced to 3 mill. (0.12 in.) in thickness, was placed in the line leading to a turn-table on the

Western Railway of Saxony, where it has remained until the present time, 1870, receiving the shocks due to engines passing to and from the turn-table more than one hundred times daily.

It follows from these experiments that the least thickness ever given to the webs of rails in practice is more than sufficient, and that if it were possible to roll webs $\frac{1}{4}$ in. thick, such webs would be amply strong, if it were not that there would be a chance of their being torn at the points where they are traversed by the fish-plate bolts. Baron von Weber concludes that webs $\frac{3}{8}$ in. or $\frac{1}{2}$ in. thick are amply strong enough for rails of any ordinary height, and that, in fact, the webs should be made as thin as the process of rolling and as the provision of sufficient bearing for the fish-plate bolts will permit.

129. ANOTHER GRAPHICAL METHOD.—If manipulating processes are to be used for determining the strength, the following method possesses many advantages over the former.

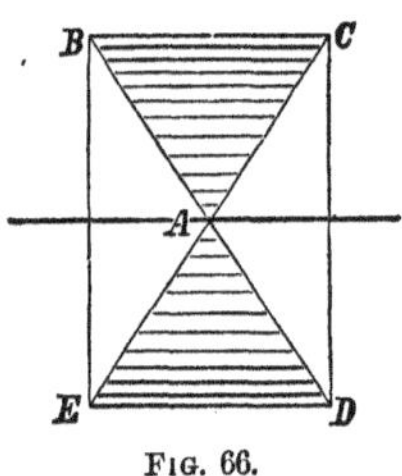

FIG. 66.

Since the strains vary directly as their distance from the neutral axis, the triangle ABC (Fig. 66), in the rectangle BCDE, represents the compressive strains if each element of the shaded part has a strain equal to R; and its moment is R times the area multiplied by the distance of the centre of gravity of the triangle from the neutral axis; or,

$$R \times (b \times \tfrac{1}{2} \text{ of } \tfrac{1}{2}d) \times \tfrac{2}{3} \text{ of } \tfrac{1}{2}d = \tfrac{1}{12}Rbd^2,$$

and the moment of tensile resistance is the same, hence the total moment is double this, or $\frac{1}{6}Rbd^2$, as found by the preceding process.

130. IF A SQUARE BEAM HAVE ONE OF ITS DIAGONALS VERTICAL (Fig. 67), the neutral axis will coincide with the other diagonal. Take any element, as ab, and project it on a line cd, which passes through A and is parallel to BC, and draw the lines Oc and Od, and note the points f and g where they intersect the line ab. If the element were at cd, the strain upon it would be R, multiplied by the area of cd, or simply R.cd; but because the strains are directly proportional to the distances of the elements from the neutral axis, the strain on ab is $R.fg$. Proceed in this way with all the elements and construct the shaded figure. The strains on the upper part of the figure ABC, which

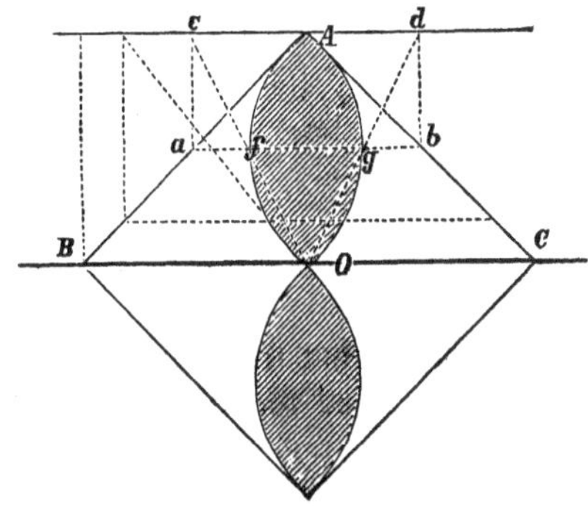

FIG. 67.

begin with zero at BC, and increase gradually to R, at A, will be equivalent to the strains on the shaded figure AO, if the strain is equal to R on each unit of its surface. Hence the total strain on each half is the area of the shaded part AO, multiplied by R, and the moment of the strain of each part is this product multiplied by the distance of the centre of the shaded part from the axis BC.

By similar triangles we have

$$Aa : ab :: AB : BC, \text{ and}$$

$$cd = ab : fg :: AO : x :: AB : Ba \text{ or } AB - Aa;$$

x being the distance of fg from O.

From these eliminate ab, and find

$$af = \tfrac{1}{2}(ab - fg) = \tfrac{1}{2}\frac{BC}{(AB)^2}(Aa)^2,$$

hence the curve which bounds the shaded figure is a parabola which is tangent to AB, and whose axis is parallel to BC.

Let d = one side of the square, then

$\sqrt{2}d = BC$,

$\frac{1}{2}\sqrt{2}d = AO$, and

$\frac{1}{4}\sqrt{2}d =$ the widest part of the shaded figure.

The area of a parabola is two-thirds the area of a circumscribed rectangle.

Hence the area of AO is

$$\tfrac{2}{3} \times \tfrac{1}{2} \sqrt{2}d \times \tfrac{1}{4} \sqrt{2}d = \tfrac{1}{6}d^2,$$

and the moment is

$$\tfrac{1}{6}d^2 \times \tfrac{1}{4} \sqrt{2}d = \frac{d^3}{12 \sqrt{2}},$$

and the moment of both sides, multiplied by R, is

$$R \frac{d^3}{6 \sqrt{2}} \quad - \; - \; - \; - \; - \; - \; - \; - \; - \; - \quad (163)$$

If $b = d$ in equation (145) and the result compared with the above, we find:—

The strength of a square beam with its side vertical : strength of the same beam with one of its diagonals vertical : : $\sqrt{2} : 1$ or as 7 : 5 nearly.

So that increased depth merely is not a sufficient guarantee of increased strength. The reason why the strength is diminished when the diagonal is vertical, is because there is a very small area at the vertex where the strain is greatest, but when a side is horizontal the whole width resists the maximum strain.

131. IRREGULAR SECTIONS.—This method is applicable to irregular sections, as shown by the following example.

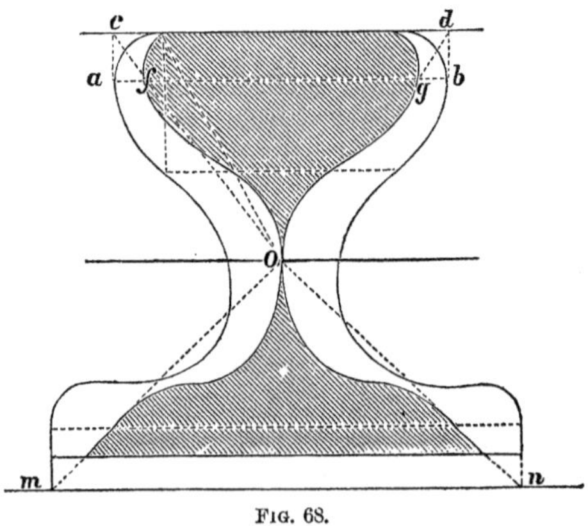

FIG. 68.

Let Fig. 68 be a cross section of a beam. In a practical case it may be well to make an exact pattern of the cross section, of stiff paper or of a thin board of uniform thickness. To find the position of the neutral axis, draw a line on the pattern which shall be perpendicular to the direction of the forces which act upon the beam, that is, if the forces are vertical the line will be horizontal. In a form like Fig. 68 this line will naturally be parallel to the base of the figure. Then balance the pattern on a knife-edge, keeping the base of the figure (or the *line* previously drawn) parallel to the knife-edge, and when it is balanced the line of support will be the neutral axis. Proceed to construct the shaded part as shown in the figure, by projecting any element, as *ab* on the line *cd*, and drawing *cO* and *dO*, and noting the intersections *f* and *g*, the same as in Fig. 67. The elements on the lower side must be projected on a line *mn*, which is at the same distance from the neutral axis as the most remote element on the upper side. The area of the shaded part above the neutral axis should equal that below, because the resistance to extension equals that for compression. The area of the shaded part may be found approximately by dividing it into small rectangles of known size, and adding together the full rectangles and estimating the sum of the fractional parts. Or, the shaded part may be cut out and carefully weighed or balanced by a rectangle of the same material, after which the sides of the rectangle may be carefully measured and contents computed. The area of the rectangle would evidently equal the area of the irregular figure.

The ordinate to the centre of gravity of each part may be determined by cutting out the shaded parts and balancing each of them separately on a knife-edge, as before explained, keeping the knife-edge parallel to the neutral axis. The distance between the line of support and the neutral axis will be the ordinate to the centre of gravity. *The moment of resistance is then found by multiplying the area of each shaded part by the distance of its centre of gravity from the neutral axis, and multiplying the sum of the products by* R.

These mechanical methods may be managed by persons who have only a very limited knowledge of mathematics, and if skilfully and carefully done will give satisfactory results. It does not, however, furnish such an *uniform*, direct and *exact* mode of solution as the analytical method which is hereafter explained.

132. FORMULA OF STRENGTH ACCORDING TO BARLOW'S THEORY.—Either of the above methods may be used. One part of the *expression* for the strength is of the same *form* as that found by the common theory; but instead of R we must use T, or C—the former if it ruptures by tension, the latter if by crushing. The other resistance, ϕ, for solid beams is evenly distributed over the surface. For example, take a rectangular beam, Fig. 61, and the resistance to *longitudinal shearing* on the upper side is $\phi\, b \times \frac{1}{2}d = \frac{1}{2}\,\phi\, bd$, and its moment is $\frac{1}{2}\,\phi\, bd \times \frac{1}{2}$ of $\frac{1}{2}d = \frac{1}{8}\,\phi\, bd^2$, and for both sides, $\frac{1}{4}\,\phi\, bd^2$. Hence, according to Barlow's theory, the expression for the strength of a rectangular beam is

$$[\tfrac{1}{4}\,\phi + \tfrac{1}{6}\mathrm{T}]\; bd^2 \text{ for cast iron, and}$$

$$[\tfrac{1}{4}\,\phi + \tfrac{1}{6}\mathrm{C}]\; bd^2 \text{ for wrought iron and wood} \qquad (164)$$

If the beam is supported at its ends and loaded at the middle, we have

$$\tfrac{1}{4}\mathrm{P}l = [\tfrac{1}{4}\,\phi + \tfrac{1}{6}\mathrm{T}]\; bd^2 \text{ for cast iron} \qquad (165)$$

The volume which represents the resistance due to ϕ is always a prism, having for its base the surface of the figure and ϕ, or some fraction of ϕ, for its altitude. If the second method of illustration be used, it will take two figures to fully illustrate the strains. For instance, if the section be as in Fig. 68, the moment of the shaded part will be multiplied by T or C, as the case may be. To find the remaining part of the moment, find the area of each part of the transverse section, also the distance of the centre of gravity of each part from the neutral axis. Then, *to find the moment of resistance due to longitudinal shearing, multiply the area of each part by the distance of its centre of gravity from the neutral axis, add the products and multiply the sum by* ϕ. This is true for solid sections; but for hollow beams, T and H sections, where there is an abrupt angular change from the flange to the vertical part of the beam, the factor ϕ requires a modification. For instance, take the simple case of a single T, Fig. 69, in which

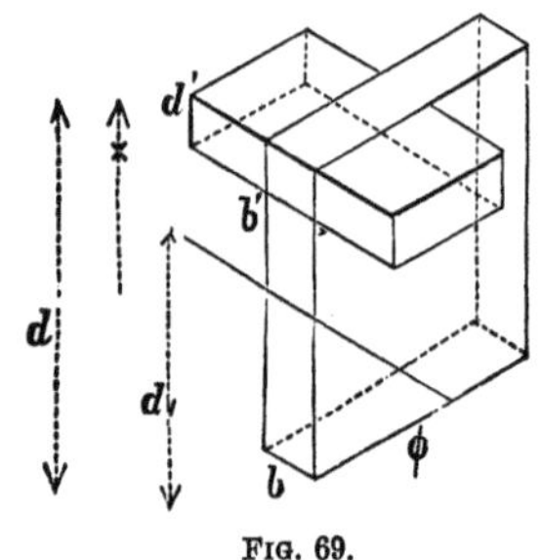

FIG. 69.

the breadth of the T is b' and its depth d', and the other notation as in the figure.

The resistance of the upper part is represented by the prism whose base is bx, and whose altitude is ϕ, *plus* the prism whose base is d' $(b'-b)$, *and whose altitude is* $\frac{d'}{x}$ ϕ. The resistance of the lower part is $\phi\, bd_1$. The total moment of this resistance is—

$$\phi\, bx.\tfrac{1}{2}x + d'(b'-b) \times \frac{d'}{x}\phi\,(x - \tfrac{1}{2}d') + \phi b d_1 \times \tfrac{1}{2}d_1$$

To this add the moment of resistance for direct extension and compression, the expression for which is of the same form as for common theory, and we have for the total moment:—

$$\tfrac{1}{2}\phi b x^2 + \frac{d'^2}{x}\phi(b'-b)(x-\tfrac{1}{2}d') + \tfrac{1}{2}\phi\; bd_1^2 + \frac{T}{3d_1}[bd_1^3 + b'x^3 - (b'-b)(x-d')^3] \quad (166)$$

From numerous experiments made upon cast-iron beams having a variety of cross sections, Barlow found that ϕ varied nearly as T, that practically it was a fraction of T, the *mean* value of which was 0.9T.

For wrought iron he found $\phi = 0.53T$

$= 0.8C$ nearly

Peter Barlow, F.R.S., father of W. H. Barlow, F.R.S., the latter of whom proposed the "theory of flexure," in an article in the Civ. Eng. Jour., Vol. xxi., p. 113, assumes that $\phi = T$.

From the above it is inferred that the practical mean values of ϕ are :—

16,000 lbs. for cast iron.
30,000 lbs. for wrought iron.
8,000 lbs. for wood.

Examples.—1. How much will a beam whose length is 12 feet, breadth 2 inches, depth 5 inches, sustain, if supported at its ends, and uniformly loaded over its whole length, and $C = 50{,}000$ lbs., $\phi = 30{,}000$ lbs., and coefficient of safety $\frac{1}{4}$? *Ans.*—11,000 lbs. nearly.

2. If $\phi = T = 16{,}000$ lbs., $b = 2$ inches, $d = 5$ inches, $l = 8$ feet; required the uniform load which it will sustain with a coefficient of safety of $\frac{1}{5}$.

3. If $b = d = 2$ inches, $l = 6$ feet, $R = 50{,}000$ lbs., is broken by an uniform load of 10,000 pounds, required ϕ.

133. BEAMS LOADED AT ANY NUMBER OF POINTS.—If the beam is loaded otherwise than has heretofore been supposed, it is only necessary to find the moment of all the forces in reference to the centre of a section and place the algebraio sum equal to the moments of resistance. Those which act in opposite directions will have contrary signs.

For instance, if a beam, AB, Fig. 70, rests upon two supports, and has weights, P_1, P_2, P_3, &c., resting upon it at distances respectively of n_1, n_2, n_3, &c., from one support, and m_1, m_2, m_3, &c., from the other, the sum of the moments of the forces on any section C whose distance is x from the support A, is

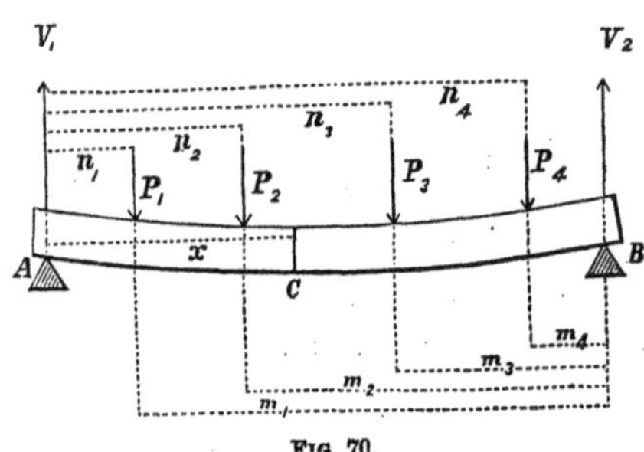

FIG. 70.

$V_1x - P_1(x - n_1) - P_2(x - n_2)$ &c., to include all'the terms of P in which n is less than x. This equals $\frac{1}{6}Rbd^2$ for rectangular beams.

V_1, the reaction of one support, is readily found by taking the moments of all the external forces about B, and solving for V_1, thus:—

$$V_1l = P_1m_1 + P_2m_2 + P_3m_3 + \&c. = \Sigma Pm$$

$$\therefore V_1 = \frac{\Sigma Pm}{l}$$

$$\text{Similarly } V_2 = \frac{\Sigma Pn}{l}$$

$$\text{also, } V_1 + V_2 = P_1 + P_2 + P_3 + \&c. = \Sigma P.$$

134. A PARTIAL UNIFORM LOAD.—Let the beam be loaded uniformly over any portion of its length, as in Fig. 71.

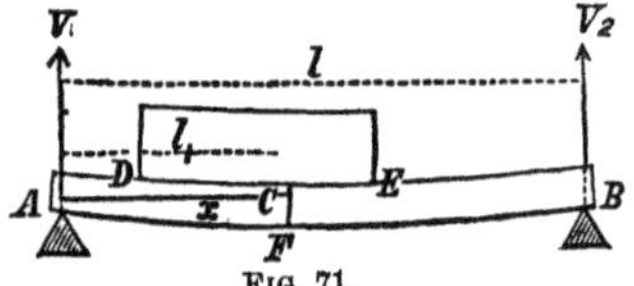

FIG. 71.

Let l = AB = length of beam;
$2a$ = DE = length of the uniform load;
x = AF = the distance to any section;
w = the load on a unit of length;
V = the reaction of the support A;
C the centre of the load;
l_1 = AC; l = CB.

Then AD $= l_1 - a$, and DF $= x - l_1 + a$.

Load on DF $= w(x - l_1 + a)$,

" " DE $= 2wa$.

By the principle of moments

$$Vl = 2wa.\, l_2 \therefore V = 2wa\,\frac{l_2}{l}.$$

The moment of stress at F is

$$Vx - \tfrac{1}{2}w(x - l_1 + a)^2$$

$$\text{or } \frac{2wal_2}{l}x - \tfrac{1}{2}w(x - l_1 + a)^2 \quad \text{- - - - - - - -} \quad (167)$$

That value of x which will make equation (167) a maximum, gives the position of the dangerous section. Differentiate, place equal zero, and make $l_1 + l_2 = l$, and solve for x, and find

$$x = a\left(1 - \frac{2l_1}{l}\right) + l_1 \quad \text{- - - - - - - - - -} \quad (168)$$

If $l_1 = \tfrac{1}{2}l,\ x = l_1$;
$l_1 < \tfrac{1}{2}l,\ x > l_1$;
$l_1 > \tfrac{1}{2}l,\ x < l_1$;
so that the maximum strain is at the centre of the loading only when the centre of the loading is over the centre of the beam; and in all other cases *it is nearer the centre of the beam than the centre of the loading is.*

The maximum strain is found by substituting the value of x equation (168) in equation (167).

The following interesting facts are also proved.

Let A D $= y \therefore a = l_1 - y$ which in equation (168) reduces it to

$$x - l_1 = (l_1 - y)\left(1 - \frac{2l_1}{l}\right) \quad \text{-} \quad (168a)$$

which is a maximum for $y = 0$; hence so far as A D is concerned, equation (168a) is a maximum when one end of the load is over the support, and for this case the equation becomes

$$x - l_1 = l_1\left(1 - \frac{2l_1}{l}\right)$$

which is a maximum for $l_1 = \tfrac{1}{4}\,l$ or $2l_1 = \tfrac{1}{2}l$, or the load must

11

extend to the middle of the beam. Making $a = l_1 = \frac{1}{4}\,l$, and equation (168) becomes

$$x = \tfrac{3}{8}\,l,$$

and these values of l_1 and x in equation (167) give for the maximum moment of stress,

$$\tfrac{9}{128} w\,l^2 = \tfrac{9}{64} W l \quad - \; - \; - \; - \; - \; - \; - \quad (169)$$

in which W is the load on half the beam.

Equation (167) gives the stress at the middle of the load, by making $a = l_1 = \frac{1}{4}\,l$ and $x = \frac{1}{4}\,l$. This gives $\frac{1}{8}\,Wl$ for the stress at the middle of the loading; hence, the maximum stress is $1\frac{1}{8}$ times the stress at the middle of the loading when the load extends from the one support to the middle of the beam.

135. OBLIQUE STRAINS.—If the force be inclined to the axis, as in Figs. 72 and 73, let θ = the angle which P makes with a normal section.

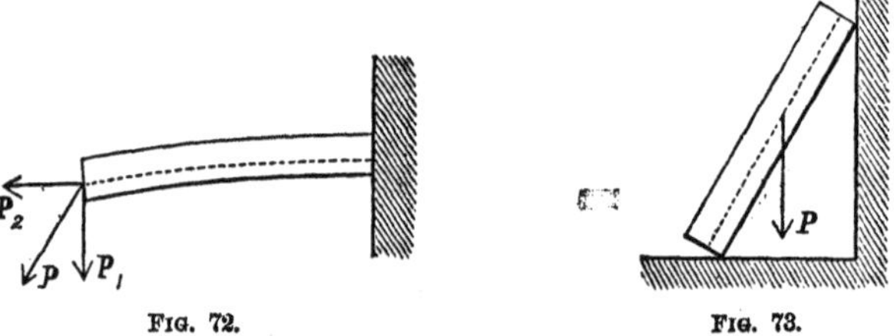

FIG. 72. FIG. 73.

Then, $P \cos\theta$ = normal component,
$P \sin\theta$ = longitudinal component.

If K = the transverse section, then

$\dfrac{P \sin\theta}{K}$ = the tension or compression upon a unit of section which arises directly from the longitudinal component. This tends directly to diminish R in the formula whether θ be obtuse or acute. If the beam be fixed at one end and free at the other, as in Fig. 72, the equation of moments becomes:—

$$P\,x \cos\theta = \left(R - \frac{P \sin\theta}{K}\right)\frac{I}{d_1}$$

which for rectangular beams becomes

$$P\, x \cos\theta = \left(R - \frac{P \sin\theta}{bd}\right)\frac{bd^2}{6} \quad - \; - \; - \; - \quad (170)$$

136. GENERAL FORMULA.—The preceding methods are easily understood, and are perhaps sufficient for the more simple cases; but for the purposes of analysis a general formula is better, by means of which a direct analytical solution may be made for special cases.

Let R = *the modulus of rupture*, as explained in article 120;
x and u horizontal coördinate axes, the former coinciding with the axis of the beam, and y a vertical axis;
Then $R\, du\, dy$ = the resistance of a fibre which is most remote from the neutral axis.
Let d_1 = distance between the neutral axis and the most remote fibre; then, according to the common theory, since the strains vary as the distance from the neutral axis $d_1 : y :: R du dy$: resistance of any fibre $= \frac{R}{d_1} y\, dy\, du$

$\therefore \frac{R}{d_1} y^2\, dy du$ = the *moment* of resistance of any fibre,

and the sum of all the moments of resistance of any section is

$$\frac{R}{d_1}\int\int y^2 dy\, du = \frac{R}{d_1} I,$$

which is called the *moment of rupture*, and must equal the sum of the moments of straining forces;

$$\therefore \Sigma P x = \frac{R}{d_1} I \quad - \; - \; - \; - \; - \; - \; - \; - \; - \; - \; - \quad (171)$$

The second member of this equation involves the character of the material (R) and the form of the transverse sections $(\frac{I}{d_1})$; the latter of which may be determined by analysis, and the former by experiment. The second member shows that for economy the material should be removed as much as possible from the neutral axis. A few special cases will now be given.

137. LET THE BEAM BE RECTANGULAR, b the breadth, and d the depth, as in Fig. 61,

$$\text{Then } I = 4\int_0^{\frac{1}{2}b}\int_0^{+\frac{1}{2}d} y^2 dy\, du = \tfrac{1}{12}bd^3$$

$$d_1 = \tfrac{1}{2}d$$

$$\therefore \frac{R}{d_1} I = \tfrac{1}{6} R\, bd^2 \text{ which is the same as expression (145).}$$

138. IF THE SIDES OF THE BEAM ARE INCLINED to the

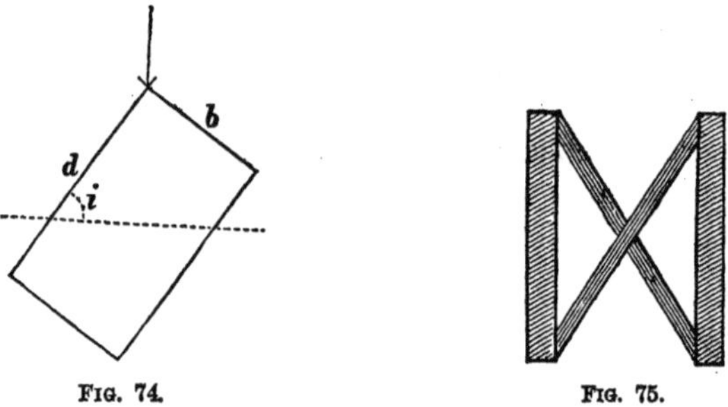

FIG. 74. FIG. 75.

direction of the force, as in Fig. 74, let i be the inclination of the side to the horizontal; then

$$I = \tfrac{1}{12}bd\,(d^2\sin^2 i + b^2\cos^2 i)^*$$

$$d_1 = \tfrac{1}{2}d\sin i + \tfrac{1}{2}b\cos i$$

$$\therefore R\frac{I}{d_1} = \tfrac{1}{6}Rbd\left[\frac{d^2\sin^2 i + b^2\cos^2 i}{d\sin i + b\cos i}\right] \quad - \; - \; - \; - \quad (172)$$

This expression has an algebraic minimum,† but not an *algebraic* maximum. By inspection, however, we find that the practical maximum is found by making $i = 90°$, if d exceeds b. Hence, a rectangular beam is strongest when its broad side is parallel to the direction of the applied forces.

Hence, the braces between joists in flooring, as in Fig. 75, not

---

* See Appendix III.

† See an article by the author in the *Journal of Franklin Institute*, Vol. LXXV., p. 260.

only serve to transmit the stresses from one to another, but also to strengthen them by keeping the sides vertical.

If $i = 90°$, equation (172) becomes $\frac{1}{6}Rbd^2$ - - - (173)

If $b = d$ and $i = 45°$, equation (172) reduces to

$$\frac{Rd^3}{6\sqrt{2}} \quad \text{- - - - - - -} \quad (174)$$

(which is the same as equation (163)),
and if $b = d$, and $i = 0°$ or $90°$, it becomes

$$\tfrac{1}{6}Rd^3.$$

Hence, the strength of a square beam having a side vertical is to the strength of the same beam having its diagonal vertical, as

$$1 : \sqrt{\tfrac{1}{2}},$$

or $\sqrt{2}$ to 1 or as 7 to 5 nearly,

In establishing equation (172) it was assumed that the *neutral surface* was perpendicular to the direction of the applied forces, which is not strictly true unless the forces coincide with the diagonal; for in other cases there is a stronger tendency to deflect sidewise than in the direction of the depth. In this case, as soon as the beam is bent there is a tendency to torsion. Both these conditions make the beam weaker than when the sides are vertical. If the tendency to torsion be neglected, the case may be easily solved; but as the result shows the advantage of keeping the sides vertical, the solution is omitted.

139. THE STRONGEST RECTANGULAR BEAM which can be cut from a cylindrical one has the breadth to the depth as 1 to $\sqrt{2}$, or nearly as 5 to 7.

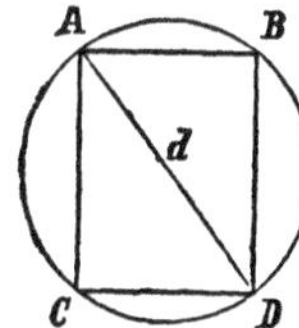

FIG. 76.

Let $x = AB =$ the breadth,
$y = AC =$ the depth, and
$D = AD =$ the diameter.

Then,

$$y^2 = D^2 - x^2$$

and equation (173) becomes

$$\tfrac{1}{6}Rxy^2 = \tfrac{1}{6}Rx(D^2 - x^2),$$

which by the Differential Calculus is found to be a maximum for

$$x = D\sqrt{\tfrac{1}{3}} \therefore y = D\sqrt{\tfrac{2}{3}}$$

$$\therefore x : y :: 1 : \sqrt{2} \text{ or nearly as 5 to 7.}$$

Examples.—How much stronger is a cylindrical beam than the strongest rectangular one which can be cut from it?

(For the strength of a cylindrical beam, see equation (180)).

Ans.—About 53 per cent.

How much stronger is the strongest rectangular beam that can be cut from a cylindrical one, than the greatest square beam which can be cut from it?

140. TRIANGULAR BEAMS.—If the base is perpendicular to the neutral axis, as in Fig. 77;

Let d = AD = the altitude, and

b = BC = the base.

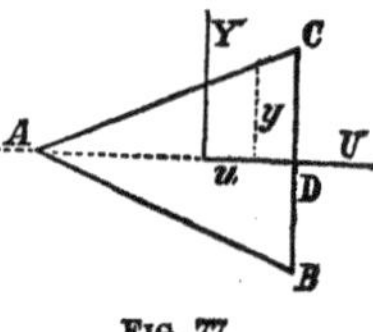

FIG. 77.

Take the origin of coördinates at the centre of gravity of the triangle, y vertical and u horizontal.

Then, by similar triangles,

$$\tfrac{1}{2}b : y :: d : \tfrac{2}{3}d + u$$

$$\therefore y = \frac{\tfrac{1}{3}db + \tfrac{1}{2}bu}{d} \therefore du = \frac{2d}{b}dy$$

$$\therefore I = 2\int\int y^2 dy du = 2\int_0^{\frac{1}{2}b} \tfrac{1}{3}y^3 du = \tfrac{1}{48}db^3.$$

We also have

$$d_1 = \tfrac{1}{2}b;$$

$$\therefore R\frac{I}{d_1} = \tfrac{1}{24}Rdb^2 = \tfrac{1}{12}RAb \quad - - - - - - - \quad (175)$$

in which A is the area of the triangle.

If the base is parallel to the neutral axis, as in Fig. 77 *a*, then, by similar triangles,

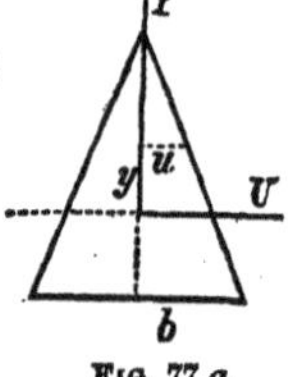

FIG. 77 *a*

$$d : \tfrac{1}{2}b :: \tfrac{2}{3}d - y : u$$

$$\therefore u = (\tfrac{2}{3}d - y)\frac{b}{2d}$$

$$\therefore \mathrm{I} = 2\int\int y^2 dy du = 2\int_{-\frac{1}{3}d}^{+\frac{2}{3}d} y^2 u dy$$

$$= \frac{b}{d}\int_{-\frac{1}{3}d}^{+\frac{2}{3}d} (\tfrac{2}{3}d - y)\, y^2 dy = \tfrac{1}{36} b d^3 \ (*)$$

We also have

$$d_1 = \tfrac{2}{3}d$$

$$\therefore \mathrm{R}\frac{\mathrm{I}}{d} = \tfrac{1}{24}\mathrm{R}bd^2 = \tfrac{1}{12}\mathrm{RA}d^2. \quad - \ - \ - \ - \ - \ - \ - \quad (176)$$

Equations (173) and (175) show that a triangular beam which has the same area and depth as a rectangular one, is only half as strong as the rectangular one.

Some authors have said that a triangular beam is twice as strong with its apex up as with it down, but this is not always the case. If the ultimate resistance of the material is the same for tension as for compression, the beam will be equally strong with the apex up or down.

If the beam is made of cast iron, and supported at its ends, it will be about 6 times as strong with the apex up as down; but if the beam be fixed at one end, and loaded at the free end, it will be about 6 times as strong with the apex down as with it up.

141. TRAPEZOIDAL BEAM.—*Required the strongest trapezoidal beam which can be cut from a given triangular one.*†

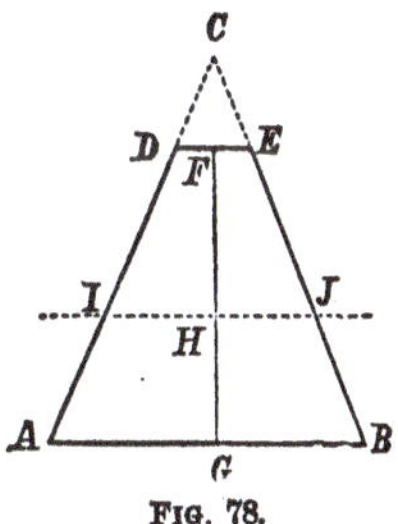

FIG. 78.

Let ABC be the given triangle,

ABED the required trapezoid,

$d = \mathrm{CG} =$ the longest altitude,

$b = \mathrm{AB}$, $d_1 = \mathrm{FH}$, $w = \mathrm{CF}$,

$z = \mathrm{CH} = d_1 + w$, and $v = \mathrm{DE}$.

IJ is the neutral axis of the trapezoid, which passes through its centre of gravity H. We may then find:—

---

* This is more easily solved by taking the moment about an axis through the vertex and parallel to the base, and using the formula of reduction. See Appendix.

† See an article by the author in the *Journal of Franklin Institute*, vol. xli., third series, p. 198.

$$d_1 = \tfrac{1}{3}\frac{d}{b} \times \frac{2b^2 - bv - v^2}{b + v}$$

$$\text{I} = \tfrac{1}{36}\frac{d^3}{b^3}\left[\frac{b^6 + b^4v - 8b^3v^2 + 8b^2v^3 - bv^4 - v^6}{b + v}\right]$$

$$\therefore \text{R}\frac{\text{I}}{d_1} = \tfrac{1}{12}\text{R}\frac{d^2}{b^2}\left[\frac{b^6 + b^4v - 8b^3v^2 + 8b^2v^3 - bv^4 - v^6}{2b^2 - bv - v^2}\right] \quad (177)$$

which is to be a maximum. By the Calculus we find, after reduction, that

$$v^3 + 5bv^2 + 7b^2v - b^3 = 0,$$

for a maximum, which solved gives

$$v = 0.13093b \text{ or } 0.13b \text{ nearly, and hence}$$
$$w = 0.13093d \text{ or } 0.13d \quad - \ - \ - \ - \quad (178)$$

which substituted in (177) gives

$$\text{R}\frac{\text{I}}{d_1} = 0.545625\,\frac{\text{R}d^2b}{12} \quad - \ - \ - \ - \ - \quad (179)$$

Dividing equation (179) by equation (176) gives 1.09125; hence from (178) and (179) we infer that *if the angle of the prism be taken off* 0.13 *of its depth, the remaining trapezoidal beam will be* 1.091 *times as strong as the triangular one, which is a gain of over* 9 *per cent.*

In order to explain this paradox it must be granted that the condition does not require that the beam shall be broken in two, but that a fibre shall not be broken—in other words, the beam shall not be fractured. The greatest strain is at the edge, where there is but a single fibre to resist it; but, after a small portion of the edge is removed, there are many fibres along the line DE, each of which will sustain an equal part of the greatest strain.

If the triangular beam were loaded so as to just commence fracturing at the edge, the load might be increased 9 per cent. and increase the fracture to only thirteen-hundredths of the depth; but if the load be increased 10 per cent. it will break the beam in two.

These results are independent of the material of which the

beam is made. If the beam be cut off $\frac{1}{3}$ the depth, its strength is found from equation (177) to be

$$0.465608 \frac{Rbd^2}{12},$$

which is 0.93101 of equation (176).

Mr. Couch found, for the mean of seven experiments on triangular oak beams of equal length, that they broke with 306 pounds. The mean of two experiments on trapezoidal oak beams, made from triangular beams of the same size as in the preceding experiments, by cutting off the edge one-third the depth when the narrow base was upward, was 284.5 pounds. This differs by less than half a pound of 0.931 times 306 pounds.

149. CYLINDRICAL BEAMS.—The moment of inertia of a circular section in which r is the radius, is

$$I = 2\iint y^2\,dy\,du = \tfrac{2}{3}\int y^3\,du = \tfrac{2}{3}\int_{-r}^{+r}(r^2-u^2)^{\frac{3}{2}}du = \tfrac{1}{4}\pi r^4$$

$$d_1 = r;$$

$$\therefore \frac{RI}{d_1} = \tfrac{1}{4}R\pi r^3 \qquad (180)$$

If polar co-ordinates are used, we have

$$du\,dy = \varrho\,d\varrho\,d\phi,$$

where ϱ is a variable radius and ϕ a variable angle.

Also $y = \varrho \sin\phi$

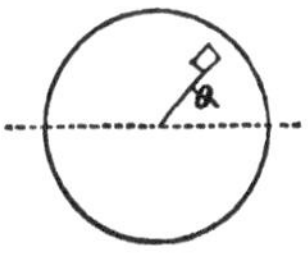

$$\therefore I = \iint y\,dy\,du = \int_0^r\int_0^{2\pi}\varrho\sin^2\phi\,d\varrho\,d\phi$$

Fig. 79.

$$= \tfrac{1}{4}r^4\int_0^{2\pi}\tfrac{1}{2}(1-\cos 2\phi)\,d\phi = \tfrac{1}{4}\pi r^4, \text{ as before.}$$

For a circular annulus we have

$$R\frac{I}{d_1} = R\frac{\pi}{4r}(r^4 - r_1^4).$$

By comparing equations (180) and (145) we see that the strength of a cylindrical beam is to that of a circumscribed rectangular one as $\frac{\pi}{32} : \frac{1}{6}$, or as 0.589 + : 1.

Also the strength of a cylindrical beam is to that of a square one of the same area as $\frac{1}{8}\text{R}\text{A}d'$ to $\frac{1}{6}\text{R}\text{A}d$ (d' being the diameter of the circle),

$$\text{or as } 1 : \left(\tfrac{4}{3}\frac{d}{d'} = \tfrac{2}{3}\sqrt{\pi}\right) \text{ or as } 1 : 1.18 \text{ nearly.}$$

143. ELLIPTICAL BEAMS.

Let $b =$ the conjugate axis, and
$d =$ the transverse axis; then
if d is vertical (Fig. 80), we have
$\text{I} = \frac{1}{64}\pi bd^3$ and $d_1 = \frac{1}{2}d$.

If b is vertical (Fig. 81), we have

$$\text{I} = \tfrac{1}{64}\pi bd^3 \text{ and } d_1 = \tfrac{1}{2}b.$$

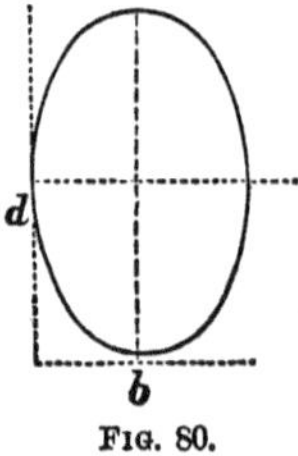

FIG. 80.

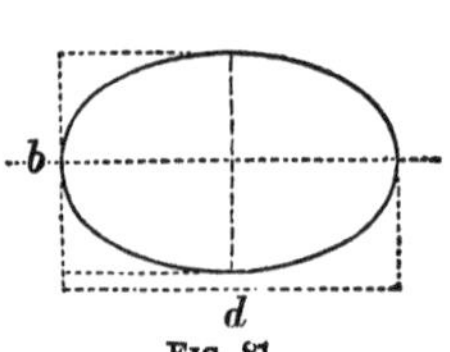

FIG. 81.

144. PARABOLIC BEAMS.

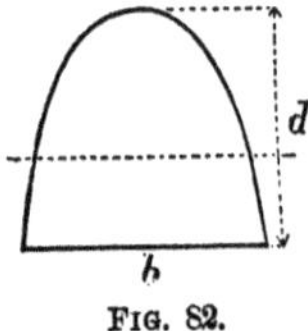

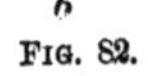

FIG. 82.

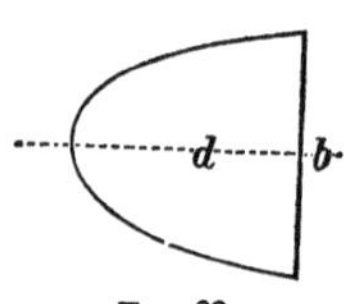

FIG. 83.

If $b =$ the base, and
$d =$ the height of the parabola, and
if d is vertical (Fig. 82), we have

$$\text{I} = \tfrac{8}{175}b^3d, \text{ and } d_1 = \tfrac{3}{5}d.$$

If b is vertical (Fig. 83), then

$$I = \frac{1}{30}bd^3, \text{ and } d_1 = \frac{1}{2}b.$$

145. According to Barlow's Theory we have

$$\frac{T}{d_1}\left[\int\int y^2 dydu\right] + \phi\int\int ydydu = \Sigma Px \quad - \quad - \quad - \quad (181)$$

which must be integrated between the proper limits to include the whole section.

If the neutral axis is at the centre of the sections, and the beam is rectangular, we have

$$\frac{T}{d_1 = \frac{1}{2}d}\left[\int_0^b\int_{-\frac{1}{2}d}^{+\frac{1}{2}d} y^2 dydu\right] + 2\phi\int_0^{+\frac{1}{2}d}\int_0^b ydydu = \Sigma Px,$$

which reduced gives

$$\tfrac{1}{6}Tbd^2 + \tfrac{1}{4}\phi bd^2 = \tfrac{1}{12}[2T + 3\phi]bd^2;$$

hence, if ϕ has any ratio to T, the *law of resistance* in solid rectangular beams is the same as for the common theory only,

If $\phi = T$, this becomes

$$\tfrac{5}{12}Tbd^2.$$

CHAPTER VII.

BEAMS OF UNIFORM RESISTANCE.

146. GENERAL EXPRESSION.—If beams are so formed that they are equally liable to break at every transverse section, they are *beams of uniform resistance*, and are generally called *beams of uniform strength.* The former term is preferable, because it applies with equal force to all strains less than that which will produce rupture. In such a beam the strain on the fibre most remote from the neutral axis is uniform throughout the whole length of the beam. The analytical condition of such a beam is: *The sum of the moments of the resisting forces must vary directly as the sum of the moments of the applied forces;* hence equation (171) is applicable; or

$$\Sigma Px = \frac{RI}{d_1}, \qquad (182)$$

which must be true for all values of x. But to obtain practical results it is necessary to consider

PARTICULAR CASES.

147. BEAMS FIXED AT ONE END AND LOADED AT THE FREE END.—*Required the form of a beam of uniform resistance when it is fixed at one end and loaded at the free end.*

1st. Let the sections be rectangular, and

$y =$ the variable depth, and
$u =$ the variable width.

Then $I = \frac{1}{12}uy^3$ (see equation (51)),
$d_1 = \frac{1}{2}y$, and
$\Sigma Px = Px =$ the variable load.*

* For ΣPx use the general moments as given in the table in Article 101, so far as they are applicable.

Hence equation (182) becomes

$$Px = \tfrac{1}{6}Ruy^2 \quad (183)$$

a. Let the breadth be constant; or $u = b$; then (183) becomes

$$Px = \tfrac{1}{6}Rby^2, \quad (184)$$

which is the equation of a parabola, whose axis is horizontal and parameter is $\frac{6P}{Rb}$. See Fig. 84.

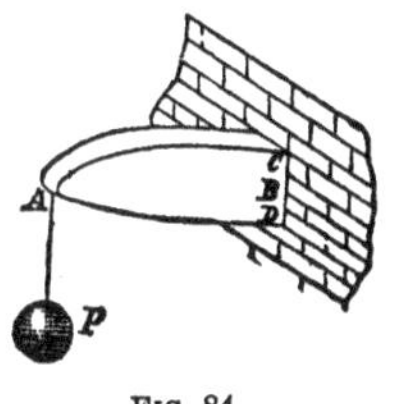

FIG. 84.

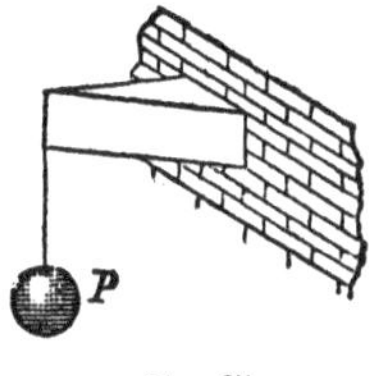

FIG. 85.

b. Suppose the depth is constant, or $y = d$. Then (183) becomes

$$Px = \tfrac{1}{6}Rd^2u, \quad (185)$$

which is the equation of a straight line; hence the beam is a wedge, as in Fig. 85.

c. If the sections are rectangular and similar, then

$$u : y :: b : d$$

$$\therefore u = \frac{b}{d}y;$$

and equation (183) becomes

$$Px = \frac{Rb}{6d}y^3,$$

which is the equation of a cubical parabola.

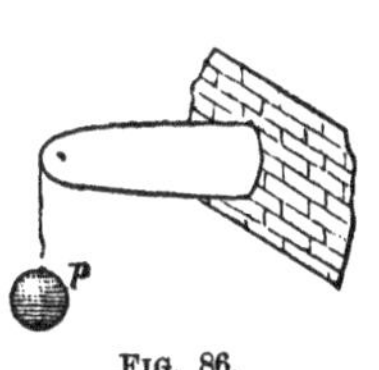

FIG. 86.

2d. Let the sections be circular. Then $I = \frac{1}{64}\pi y^4$ (equation (52), in which y is the diameter of the circle), and $d_1 = \frac{1}{2}y$; hence (182) becomes

$$Px = \tfrac{1}{32}R\pi y^3,$$

which is also the equation of a cubical parabola, as shown in Fig. 86.

3d. Let the transverse sections be rectangular, and I constant, the breadth and depth both being variable, then equation (182) becomes

$$Px = R\frac{(\frac{1}{12}uy^3)}{\frac{1}{2}y} = R\frac{c}{6y} \quad - \; - \; - \; - \; - \; - \quad (186)$$

in which c is a constant, $= bd^3$, b and d being the breadth and depth at the fixed end. Equation (186) is the equation of the vertical longitudinal sections, and is the equation of an hyperbola

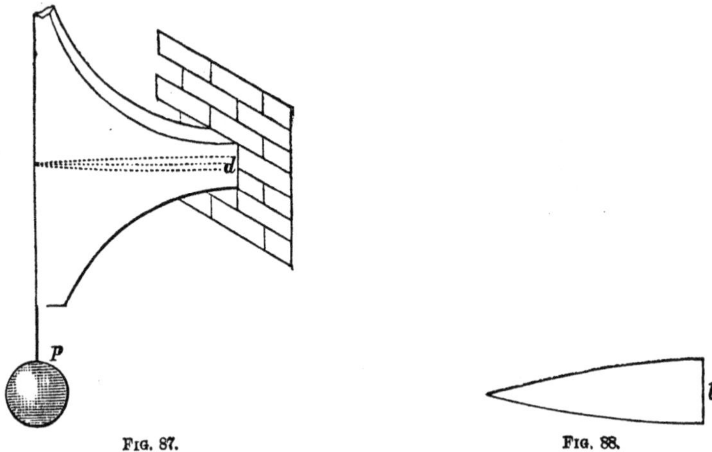

FIG. 87. FIG. 88.

referred to its asymptotes. See Fig. 87. If the value of y from this equation be substituted in the equation $uy^3 = c$, it gives

$$u = \frac{216P^3x^3}{b^2d^6} \quad - \; - \; - \; - \; - \; - \; - \; - \; - \quad (187)$$

which is the equation of the horizontal longitudinal sections; hence they are cubical parabolas, as in Fig. 88. For x and $u = 0$, $y = \infty$, and for $x = l$, $u = b = \frac{216P^3l}{b^2d^6}$

4th. If the breadth is the nth power of the depth, and the sections are rectangular, then $u = y^n$, and equation (183) becomes

$$Px = \tfrac{1}{6}Ruy^2 = \tfrac{1}{6}Ry^{n+2},$$

which is the general equation of parabolas.

148. BEAMS FIXED AT ONE END AND UNIFORMLY LOADED.—*Required the form of a beam of uniform resistance* *when it is fixed at one end and uniformly loaded over its whole length; the weight of the beam being neglected.*

The origin of co-ordinates being still at the free end, we have

$wx =$ the load on a length x, and

$\frac{1}{2}wx^2 =$ the moment of the load (equation (53)).

Hence, for rectangular sections, equation (182) becomes

$$\tfrac{1}{2}wx^2 = \tfrac{1}{6}Ruy^2 \qquad (188)$$

a. If the breadth is constant, or $u = b$ in (188), it becomes

$$\tfrac{1}{2}wx^2 = \tfrac{1}{6}Rby^2,$$

which is the equation of a straight line; and hence the beam will be a wedge, as in Fig. 89.

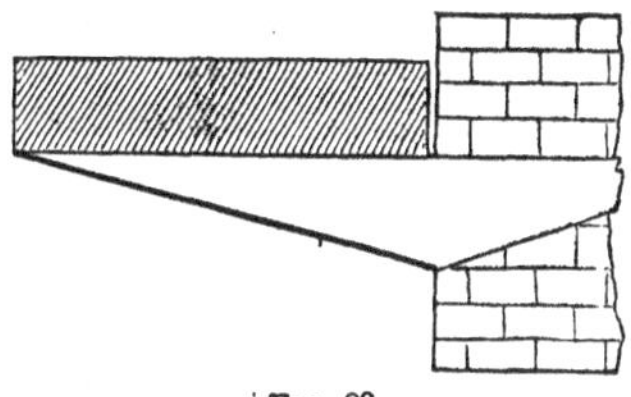

FIG. 89.

b. Let the depth be constant; or $y = d$ in (188)

$$\therefore \tfrac{1}{2}wx^2 = \tfrac{1}{6}Rd^2u;—$$

a parabola whose axis is perpendicular to the axis of the beam, as in Fig. 90.

c. Let the sections be similar;—

then $d : b :: y : u = \frac{b}{d}y,$

$\therefore$ equation (188) becomes $\frac{1}{2}wx^2 = \frac{1}{6}R\frac{b}{d}y^3;$—

a semi-cubical parabola, as in Fig. 91.

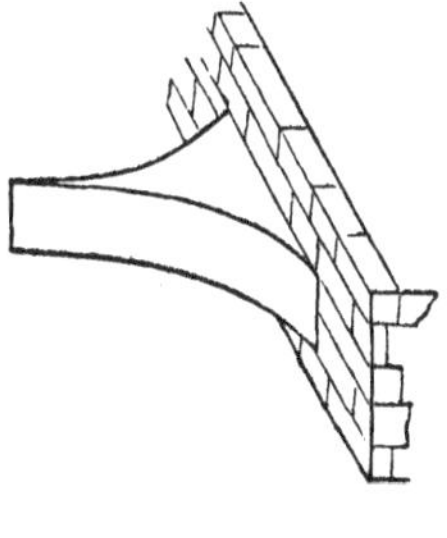

FIG. 90.

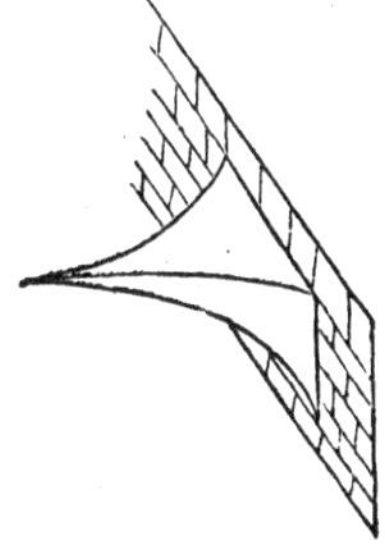

FIG. 91.

d. Let I be constant, or $\frac{1}{12}uy^3 = \frac{1}{12}bd^3$. Then equation (182) becomes

$$\tfrac{1}{2}wx^2 = \tfrac{1}{6}\text{R}\,\frac{bd^3}{y};\text{—an hyperbola of the second order.}$$

149. PREVIOUS CASES COMBINED.—*Required the form of the beam of uniform resistance when it is fixed at one end and loaded uniformly, and also loaded at the free end.*

The moment of applied forces is $\text{P}x + \frac{1}{2}wx^2$; hence equation (182) becomes, for rectangular beams,

$$\text{P}x + \tfrac{1}{2}wx^2 = \tfrac{1}{6}\text{R}uy^2.$$

Hence, if the depth is constant, $\text{P}x + \frac{1}{2}wx^2 = \frac{1}{6}\text{R}ud^2$;—a parabola;

Hence, if the breadth is constant, $\text{P}x + \frac{1}{2}wx^2 = \frac{1}{6}\text{R}by^2$;—an ellipse;

Hence, if the sections are similar, $\text{P}x + \frac{1}{2}wx^2 = \frac{1}{6}\text{R}\frac{b}{d}y^3$;—a semi-cubical parabola.

150. WEIGHT OF THE BEAM CONSIDERED.—*Required the form of the beam of uniform resistance when the weight of the beam is the only load; the beam being fixed at one end and free at the other.*

a. Let the sections be rectangular and the breadth constant.

Let $x = \text{AB}$; Fig. 92,

$y = \text{DC}$,

$b =$ the breadth, and

$\delta =$ the weight of a unit of volume.

Then $\int ydx =$ the area of ADC, and

$\delta b\int ydx =$ the weight of ADC;

the limits of the integration being 0 and x.

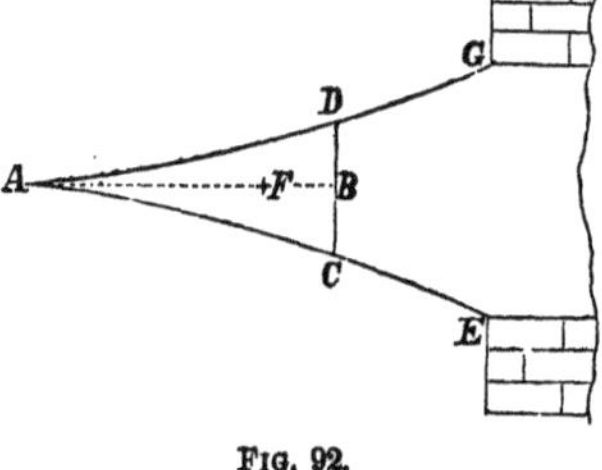

FIG. 92.

If F is the centre of gravity of ADC; we have, from the principles of mechanics, the distance $\text{AF} = \frac{\int xydx}{\int ydx}$.

The moment of the applied forces is the weight of ADC multiplied by the distance $\text{BF} = x - \text{AF}$. Hence, equation (182) becomes

$$\delta b \int y dx \left[x - \frac{\int xy dx}{\int y dx} \right] = \tfrac{1}{6} b y^2,$$

which reduced gives

$$x^2 = \frac{2\text{R}}{\delta} y \quad - \; - \; - \; - \; - \; - \; - \; - \; - \; - \; - \; - \; - \; - \quad (189)$$

which is the equation of the common parabola, the axis being vertical.

If there is a single curve, $\frac{2\text{R}}{\delta}$ is its parameter; but if two curves, as in the figure, $\frac{\text{R}}{\delta}$ is the parameter of each.

b. Let the depth be constant. In a similar way we find

$$\delta d \int u dx \left[x - \frac{\int ux dx}{\int u dx} \right] = \tfrac{1}{6} \text{R} d^2 u.$$

This solved gives

$$x = -\sqrt{\frac{\text{R}d}{6u}} \text{ Nap. log.} \left[\sqrt{\frac{\text{R}d}{6u} C + u^2} - u \right] + C',$$

in which C and C' are constants of integration, and involve the position of the origin of co-ordinates and direction of the curve at a known point.

c. Let the beam be a conoid of revolution, as in Fig. 93.

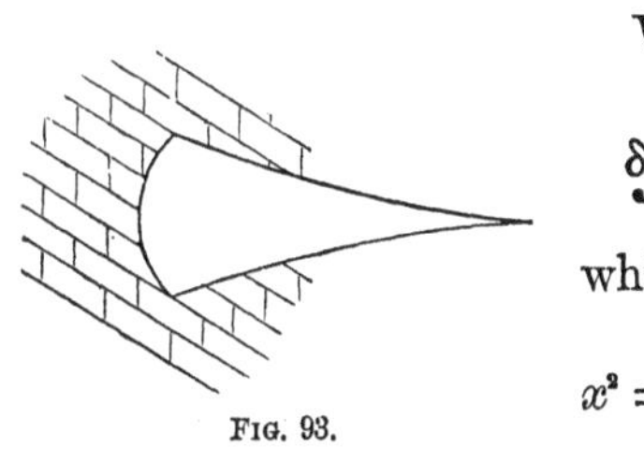

FIG. 93.

We have, as before

$$\delta \int \pi y^2 dx \left[x - \frac{\int \pi y^2 x dx}{\int \pi y^2 dx} \right] = \tfrac{1}{4} \pi \text{R} y^3,$$

which reduced gives

$$x^2 = \tfrac{15}{2} \frac{\text{R}}{\delta} y \quad - \; - \; - \; - \; - \; - \; - \quad (191)$$

which is the equation of the common parabola.

d. Suppose, in the preceding cases, that an additional load, P, *is applied at the free end.*

Some of the equations which result from this condition cannot be integrated in finite terms, and hence the curves cannot be classified.

151. BEAMS SUPPORTED AT THEIR ENDS.

A. *Let the beam be supported at its ends and loaded at the middle point.*

For this case, equation (182) becomes, for rectangular sections,

$$\tfrac{1}{4}\mathrm{P}x = \tfrac{1}{6}\mathrm{R}uy^2 \qquad (192)$$

a. If the breadth is constant, we have

$$\tfrac{1}{4}\mathrm{P}x = \tfrac{1}{6}\mathrm{R}by^2,$$

which is the equation of the common parabola.

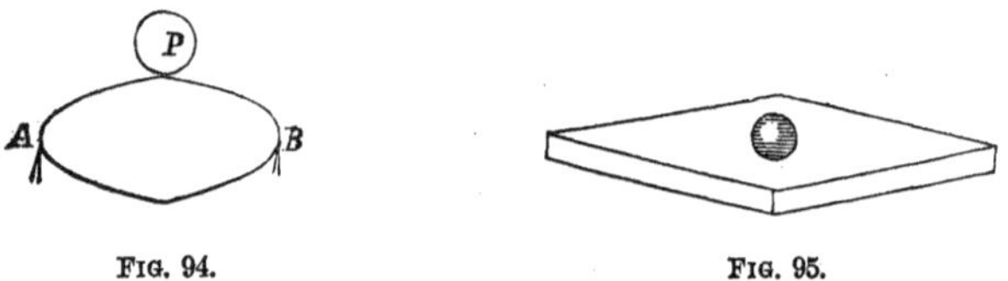

FIG. 94. FIG. 95.

The beam consists of two parabolas, having their vertices, one at each support, as in Fig. 94.

b. If the depth is constant, we have

$$\tfrac{1}{4}\mathrm{P}x = \tfrac{1}{6}\mathrm{R}d^2y; \qquad (193)$$

a wedge, as in Fig. 95.

B. *If the beam is uniformly loaded*, we have from equations (74) and (182),

$\tfrac{1}{2}w\ (lx - x^2) = \tfrac{1}{6}\mathrm{R}uy^2$—if rectangular, and if the breadth is constant, $\tfrac{1}{2}w\ (lx - x^2) = \tfrac{1}{6}\mathrm{R}by^2$; - - - - - - - (194) an ellipse, Fig. 96.

If the depth is constant, $\tfrac{1}{2}w\ (lx - x^2) = \tfrac{1}{6}\mathrm{R}d^2u$, a parabola, Fig. 97.

FIG. 96. FIG. 97.

C. *Let the beam have an uniform load and also an uniformly increasing load from one end to the other, as in* Fig. 98.

Let W = the weight of the uniform load,

W_1 = the weight of the uniformly increasing load, and

V = the reaction of the support at the end which has the least load.

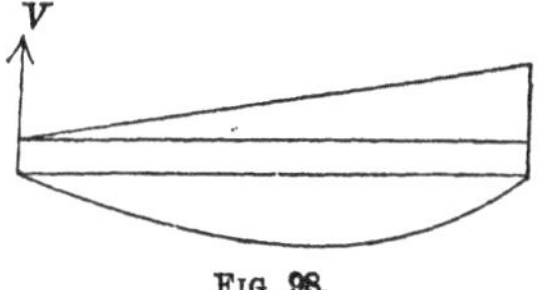

FIG. 98.

Then $V = \frac{1}{2}W + \frac{1}{3}W_1$.

Let x be reckoned from A, then the load on x is

$$\frac{W}{l}x + \frac{W_1}{l^2}x^2,$$

and the moment of this reaction and load on a section which is at a distance x from A is

$$(\tfrac{1}{2}W + \tfrac{1}{3}W_1)x - \frac{Wx^2}{2l} - \frac{W_1x^3}{3l^2} \quad - \; - \; - \; - \; - \; - \quad (195)$$

which equals $\frac{1}{6}R\,by^2$ for rectangular beams of uniform breadth. To find the point of greatest strain, make the first differential coefficient of (195), equal to zero. We thus find

$$\tfrac{1}{2}W + \tfrac{1}{3}W_1 - \frac{W}{l}x - \frac{W_1}{l^2}x^2 = 0.$$

If W = 0, this gives

$$x = \tfrac{1}{3}l\sqrt{3}.$$

When W = 0, this becomes the case of water pressing against a vertical surface.

152. BEAMS FIXED AT THEIR ENDS.—If the beam is fixed at its ends and loaded at the middle with a weight, P, we have, from equations (117) and (182), when the breadth is uniform,

$$\tfrac{1}{8}P(l - 4x) = \tfrac{1}{6}R\,by^2, \quad - \; - \; - \; - \; - \; - \; - \quad (196)$$

which is the equation of a parabola. The beam really consists of four double parabolas with their vertices tangent to each other, as in Fig. 99. The vertices are $\frac{1}{4}l$ from the end.

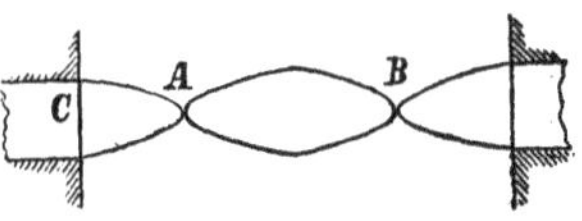

FIG. 99.

If the load were uniform we would obtain, in a similar way, a beam composed of four wedges. These are direct deductions from theory, but it is evident that there is something wanting, for a beam like Fig. 99 has no transverse strength. The same result, though not quite so glaringly apparent at first sight, exists in all the cases which we have discussed. For instance, in figures 85, 86, and 87 the sections at the free end must have a finite value to resist the shearing stress, and the beams must be enlarged, as determined in the next article. If the section is reduced to naught, it can sustain no weight. In the present case, there is neither tension nor compression at A and B, as was shown in articles 99 and 100; but there is a *transverse shearing* stress at those points, and there must be sufficient transverse section to resist it. The same remark applies to the preceding cases, and the forms must all be modified to meet this condition, as is shown in the next article.

153. EFFECT OF TRANSVERSE SHEARING STRESS *on modifying the forms of the beams of uniform resistance.*

The value of the transverse shearing stress is given in Article 84. For instance, in the case of a beam uniformly loaded, it is $V - wx = \frac{1}{2}wl - wx = \frac{1}{2}w(l - 2x)$ at any point in the length. This quantity, divided by the product of the breadth and *modulus of strength for transverse shearing*, gives the depth necessary to sustain this force. Take, for example, case A, Article 134. The load being uniform, we have $Ss = \frac{1}{2}w(l - 2x)$ as given above, which is the equation of a straight line, Fig. 100, in which

FIG. 100. FIG. 101.

$$AB = \tfrac{1}{2}wl \div (b \times \textit{modulus of shearing}).$$

Hence we would at first thought naturally infer that the form of the beam of uniform strength in this case would be found by adding the ordinates of the straight line, AC, to the corresponding ordinates of the ellipse, thus giving Fig. 101. But as soon as this is done the equation of moments is changed; for the lever arm of the force is increased, and the moment of resisting forces is greater. To avoid this difficulty we would add the section which is necessary for sustaining the shearing stress, to the side of the beam. But in all those cases where the depth, as found by moments, is zero, this method is impracticable, for the thickness to be added would be infinite. *It seems, then, that to solve this case theoretically, we must add some arbitrary quantity to the depth as found by moments, which quantity shall increase the section so as to fully resist the moment of the applied forces, and, in addition thereto,* PARTLY *resist the shearing stress, and then a section must be added to the side of the beam which shall sustain the remainder of the shearing stress.*

Tabulated values of shearing stresses for several of the cases which have been considered. The values in the fourth column of the following table may be found according to the principles given in Article 84, or they may be found by taking the first differential coefficient of the moments of applied forces.*

---

* The third of equations (42*a*) is $\Sigma Px - \Sigma F \sin. a \times x = \Sigma F_1 y$, and since the lever arms, x and y, of the forces are always linear quantities, we may enter under the sign Σ and differentiate them. This gives $\Sigma P dx - \Sigma F \sin a dx = (\Sigma F_1) dy$, or $\frac{dx}{dy}[\Sigma P - \Sigma F \sin a] = \Sigma F$, which, combined with the second of (42*a*), gives $\Sigma P = \Sigma F \sin a = Ss$. Hence we have this simple rule: Ss *is the first differential coefficient of the moments of the applied forces.*

When the bending moment has an algebraic maximum, the abscissa of the point of greatest bending stress may be found by making the first differential coefficient of the moment of the stress equal to zero, and solving for x; hence, *in this case*, the bending moment is greatest where the shearing stress is *zero*.

A TABLE OF MOMENTS AND SHEARING STRESS.

Number of the Case. See page 130.	Condition of the beam.	General moments of the Applied Forces. See page 130.	Shearing Stress. S_s.
I.	Fixed at one end and P at the free end.	Px. Eq. (35).	P.
II.	Fixed at one end and uniformly loaded.	$\frac{1}{2}wx^2$. Eq. (40).	wx.
IV.	Supported at the ends and P at the middle.	$\frac{1}{2}Px$. Eq. (44).	$\frac{1}{2}$P.
V.	Supported at the ends and uniformly loaded.	$\frac{1}{2}wlx-\frac{1}{2}wx^2$. Eq. (48).	$\frac{1}{2}wl-wx$.
VII.	Fixed at one end, supported at the other, and P anywhere.	For the part AD, Fig. 43, Eq. (64). For the part DB, Fig. 43, Eq. (67).	$(-1+\frac{1}{2}n^2(3-n))$P. $\frac{1}{2}n^2(3-n)$P.
VIII.	Fixed at one end, supported at the other, and uniformly loaded.	$\frac{1}{8}w(4x^2-3lx)$. Eq. (87).	$\frac{1}{8}w(8x-3l)$.
IX.	Fixed at the ends and P at the middle.	$\frac{1}{8}(l-4x)$P. Eq. (94).	$\frac{1}{2}$P.
X.	Fixed at the ends and uniformly loaded.	$\frac{w}{12}(l^2-6lx+6x^2)$. Eq. (102).	$\frac{1}{2}wl-wx$.
Art. 150.	Fixed at one end, and the weight of the beam the load.	$\delta bx\int ydx-\delta b\int xydx$.	$\delta b\int ydx$.
C. Art. 151.	Supported at the ends, uniform load, also load uniformly increasing.	$(\frac{1}{2}W+\frac{1}{3}W_1)x-\frac{Wx^2}{l^2}-\frac{W_1}{3l^2}x^3$.	$\frac{1}{2}W+\frac{1}{3}W_1-\frac{Wx}{l}-\frac{W_1x^2}{l^2}$.

If a beam is supported at its ends, and loaded with several weights P_1, P_2, P_3, etc., as in Fig. 102, we may readily find the shearing stress at any point by article 84. It is there shown that the *shearing stress* $= \Sigma P$, where ΣP equals the algebraic sum of all the vertical forces, including the reaction at the abutment. Hence, we have for the *shearing stress*

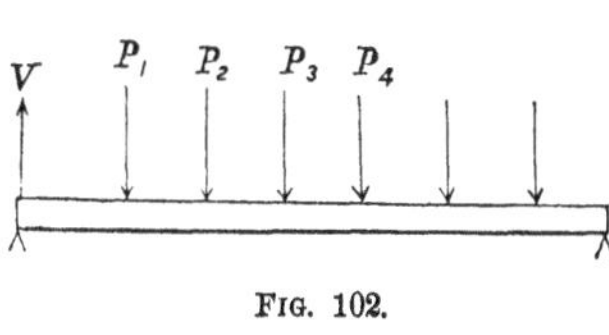

FIG. 102.

between the end and $P_1 = V$;
between P_1 and P_2 $= V - P_1$;
between P_2 and P_3 $= V - P_1 - P_2$;
between P_3 and P_4 $= V - P_1 - P_2 - P_3$; etc.

If the weights are equal to each other $= P$, we have $P = P_1 = P_2 = P_3$, etc.; and if there are n of them, and they are symmetrically placed in reference to the centre of the beam, we have

$$V = \tfrac{1}{2}nP.$$

If n is even, we have, at the centre of the beam, the

transverse *shearing stress* $= \frac{1}{2}nP - \frac{1}{2}nP = 0$ - - (197);

and if n is odd, there will be a weight at the centre, and each side of the central weight we have

transverse *shearing stress* $= \frac{1}{2}nP - \frac{1}{2}(n \pm 1)P = \pm \frac{1}{2}P$ - (198).

154. UNSOLVED PROBLEMS.—Many practical problems in regard to the resistance of materials cannot be solved according to any known laws of resistance. Some of these have been solved experimentally, and empirical formulas have been deduced from the results of the experiments, which are sufficiently exact for practical purposes, within the range of the experiments. The resistance of tubes to collapsing, the strength of columns, and the proper thickness of the vertical web of rails, are such problems which have been solved experimentally. The following problems are of this class, and have not been solved. The first four are taken from the *Mathematical Monthly*, Vol. I., page 148.

1. Required a formula for the strength of a circular flat iron

plate of uniform thickness, supported throughout its circumference and loaded uniformly.

2. Required the strength of the same plate if the edges are bolted down.

3. Required the equation of the curve for each of the preceding cases, that they may have the greatest strength with a given amount of material.

4. In the preceding problems, suppose that the plate is square.

5. Required the form of a beam of uniform strength which is supported at its ends, the weight of the beam being the only load. Suppose, also, that it is loaded at the middle.

The latter part of this problem has received an approximate solution under certain conditions, as will be seen from the following experiments.

155. BEST FORM OF CAST-IRON BEAM AS FOUND EXPERIMENTALLY.—Cast-iron beams were first successfully used for building purposes by Messrs. Boulton and Watt. The form of the cross-section of the beams which they used is shown in Fig. 103. More recent experiments show that this is a good form, but not the best.

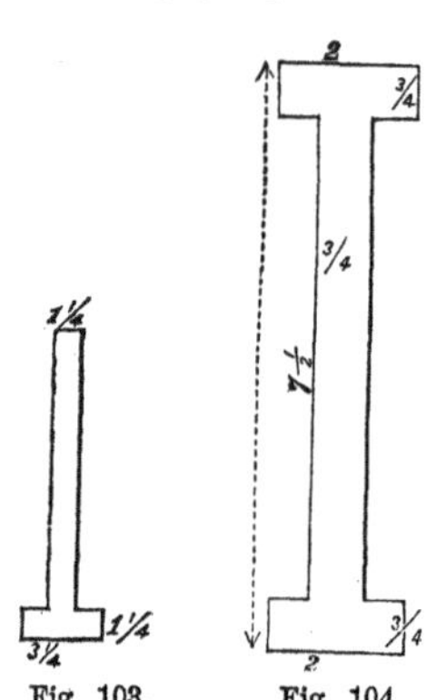

Fig. 103. Fig. 104.

About 1822 Mr. Tredgold made an experiment upon a cast-iron beam of the form shown in Fig. 104, to determine its deflection. He recommended this form for beams.

Mr. Fairbairn has justly the credit of making the first series of experiments for determining the best *form of the beam*. These experiments were prosecuted by himself for a few years, beginning about 1822, and continued still later by Mr. Hodgkinson.

The experiments quickly indicated that the lower flange should be considerably the largest.

The following experiments were made by Mr. Hodgkinson (Fairbairn on Cast and Wrought Iron, p. 11).

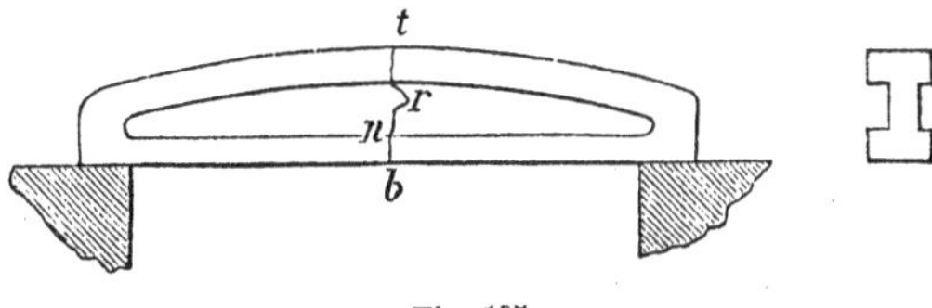

Fig. 105.

Fig. 105 shows the elevation and cross-section of a beam whose dimensions are as follows :—

Area of top rib = 1.75 × 0.42 = 0.735 inches.
Area of bottom rib = 1.77 × 0.39 = 0.690 "
Thickness of vertical rib, - - 0.29 "
Depth of the beam, - - - - - - 5.125 "
Distance between the supports, 54.00 "
Area of the whole section, - - 2.82 square inches.
Weight of the beam, - - - - 36¼ pounds.
Breaking weight, - - - - - - 6,678 pounds.

The form of the fracture is shown at *b n r*. It broke by tension.

Experiment IV.

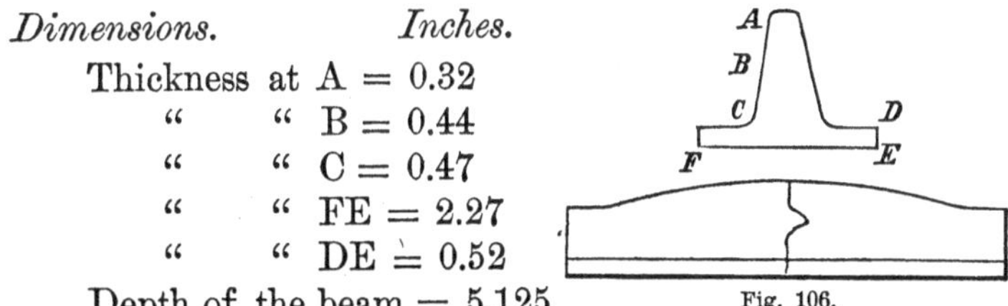

Dimensions. *Inches.*

Thickness at A = 0.32
" " B = 0.44
" " C = 0.47
" " FE = 2.27
" " DE = 0.52

Depth of the beam = 5.125

Fig. 106.

Area of the section = 3.2 square inches.
Distance between the supports = 54 inches.

Weight of casting = 40½ lbs.
Deflection with 5,758 lbs. = 0.25 inches.
" " 7,138 " = 0.37 "
Breaking weight 8,270, lbs.

EXPERIMENT 19.

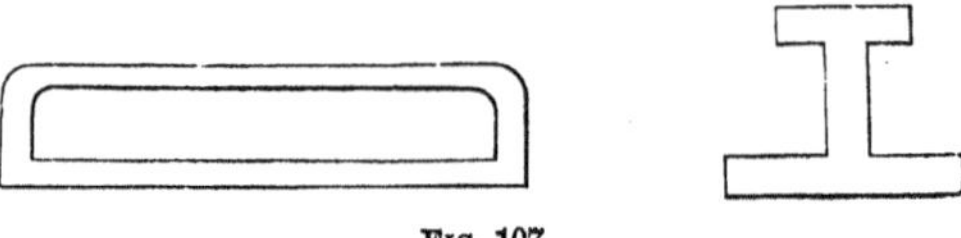

FIG. 107.

Dimensions in inches:—
Area of top rib = 2.33 × 0.31 = 0.72.
" " bottom rib = 6.67 × 0.66 = 4.4.
Ratio of the area of the ribs = 6 to 1.
Thickness of vertical part = 0.266.
Area of section, 6.4.
Depth of beam, $5\frac{1}{8}$.
Distance between the supports, 54 inches.
Weight of beam, 71 lbs.

This beam broke by compression at the middle of the length with 26,084 lbs.

It is probable that the neutral was very near the vertex *n*, or about $\frac{3}{4}$ the depth.

EXPERIMENT 21.

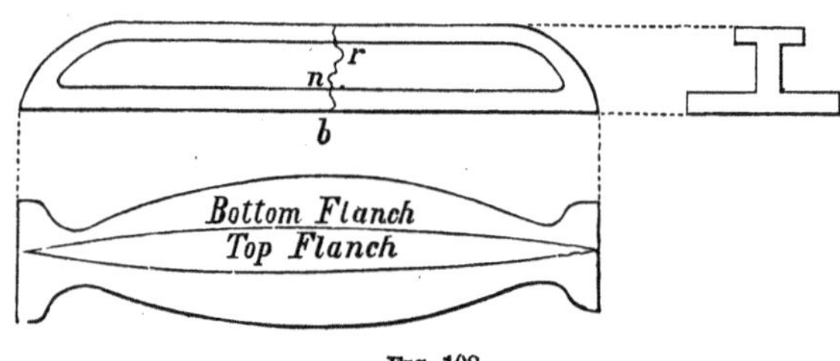

FIG. 108.

This was an elliptical beam, Fig. 108.
Dimensions in inches:—
Area of top rib = 1.54 × 0.32 = 0.493.
" " bottom rib = 6.50 × 0.51 = 3.315.
Ratio of ribs, $6\frac{1}{2}$ to 1.
Thickness of vertical part = 0.34.
Depth of beam, $5\frac{1}{8}$.
Area of the section, 5.41.
Distance between supports, 54 inches.
Weight of beam, $70\frac{3}{4}$ lbs.

Broke at the middle by tension with 21,009 lbs.

Form of fracture $b\,n\,r$; $b\,n = 1.8$ inches.

As these beams have all the same depth and rested on the same supports, 4 feet 6 inches apart, their relative strengths will be *approximately* as the breaking weight divided by the area of the cross section.

In Experiment 1, 6,678 ÷ 2.82 = 2,368 lbs. per square inch.
" " 14, 8,270 ÷ 3.2 = 2,584 " "
" " 19, 26,084 ÷ 6.4 = 4,075 " "
" " 21, 21,009 ÷ 5.41 = 3,883 " "

It is evident from these experiments, that when the vertical rib is thin, the area of the lower rib should be about 6 times that of the upper. In the 19th experiment it has already been observed that the beam broke at the top, and in the 21st it broke at the bottom, although the lower flange was larger in proportion to the upper than in the preceding case, and the comparison shows that they were about equally well proportioned. They should be so proportioned that they are equally liable to break at the top and bottom.

A beam proportioned so as to be similar to either of the two last forms above mentioned may be called a "TYPE FORM."

156. HODGKINSON'S FORMULAS *for the strength of cast-iron beams of the* TYPE FORM.

Let $W =$ the breaking weight in tons (gross).
$a =$ the area of bottom rib at the middle of the beam.
$d =$ the depth of the beam at the middle.
and $l =$ the distance between the supports.

Then according to Mr. Hodgkinson's experiments we have

$W = 26\dfrac{ad}{l}$ when the beam was cast with the bottom rib up, and

$W = 24\dfrac{ad}{l}$ when the beam is cast on its side.

157. EXPERIMENTS ON T RAILS.—Experiments on T bars, supported at their ends and loaded at the middle, gave the following results:—*

* Mahan's Civ. Eng., pp. 88 and 89; Barlow on the Strength of Materials, p. 183.

Hot blast bar, rib upward, ⊥ broke with - - 1,120 pounds.
" " " downward, T broke with - 364 "
Cold blast " " upward, ⊥ broke with. - - 2,352 "
" " " downward, T broke with - - 980 "

The ratio of the strengths is nearly as 3 to 1, but according to the table in Article 47, we might reasonably expect a higher ratio. If a greater number of experiments would not have given a higher ratio, we would account for the discrepancy by supposing that the neutral axis moved before rupture took place, or that the ratio of the crushing strength and tenacity is less for comparatively thin castings than for thick ones. It is known that the crushing strength of thin castings is proportionately stronger than thick ones. Hodgkinson found that for castings 2, 2½, and 3 inches thick, the *crushing strengths* were as 1 to 0.780 to 0.756 ; and Colonel James found a greater increase—being as 1 to 0.794 to 0.624. See also Article 37.

158. WROUGHT-IRON BEAMS.—The treacherous character of cast-iron beams, on account of the internal structure of the metal, and the unseen cracks and flaws which may exist, has led to the introduction of solid wrought-iron beams. When cast-iron beams were first used, it was practically impossible to manufacture solid wrought-iron ones, but the great improvements which have been made since then in the processes of manufacturing, have not only made their construction possible, but they have enabled the manufacturer to produce them so cheaply as to bring them within the means of those who desire such articles. At Trenton and Pittsburg they make rolled beams from a single pile,* but it is stated that by this method they can make beams only about nine inches in depth. At Buffalo and Phœnixville they use Mr. John Griffin's patent, which consists in rolling the flanges separately, piling the plates for the web between them, and then rolling and welding the whole together. By this method they can make beams at least twenty inches deep, and of any desired length. There is no attempt to make them of uniform strength. They are of the double T (I) pattern, and of uniform section throughout.

* Jour. Frank. Inst., Vol. 86, p. 231.

CHAPTER VIII.

TORSION.

159. TORSIVE STRAINS are very common in machinery. In all cases where a force is applied at one point of a shaft to turn (or twist) it, and there is a resisting force at some other point, the shaft is subjected to a torsive strain. The wheel and axle is a familiar case in which the axle is subjected to this strain. To produce torsion without bending, two equal and parallel forces, acting in opposite directions, and lying in a plane which is perpendicular to the axis of the piece, must be so applied to the section that the arms of the forces shall be equal. In other words, mechanically speaking, a *couple* whose axis coincides with the axis of the piece, must be applied to the piece. If only a single force, P, is applied, as in Fig. 109, the piece is pushed sidewise at the same time that it is twisted; but the amount of twisting is the same as if the force, P, were divided into two, each equal $\frac{1}{2}P$, and each of these acted on opposite sides of the axis and in opposite directions, and at a distance from the axis equal AB, Fig. 109. For, the moment of the couple thus formed, is $\frac{1}{2}P \times 2 \times AB = P.AB$, which is the moment of P.

160. THE ANGLE OF TORSION *is the angle through which a fibre whose length is unity, and which is situated at a unit's distance from the axis, is turned by the twisting force.* It depends for its value, in any case, upon the elastic resistance to torsion, as well as upon the dimensions of the piece and the twisting force. The analysis by which its value is determined is founded upon the following hypotheses, which are approximately correct.

1st. The resistance of any fibre to torsion varies directly as its distance from the axis of the piece.

2d. The angular amount of torsion of any fibre between any two sections, or the total angle of torsion, varies directly as the distance between them.

It is found by experiment, that these hypotheses are sufficiently exact for cylinders and regular polygonal prisms of many sides. They assume that transverse sections which were plane before twisting, remain so while the piece is twisted, but in reality the fibres which were parallel to the axis before being twisted are changed to helices, and this operation produces a longitudinal strain upon the fibres; and this, in turn, changes the transverse sections into warped surfaces.*

To find the angle of torsion:—

Let $l =$ AD $=$ the length of the piece, Fig. 109.

$a =$ AB $=$ the lever arm of P.

P $=$ the twisting force.

$\alpha = a\text{A}b =$ the *total angle of torsion*, or angle through which Aa has been twisted.

$\theta = \frac{\alpha}{l} =$ "The Angle of Torsion,"

—supposed to be small.

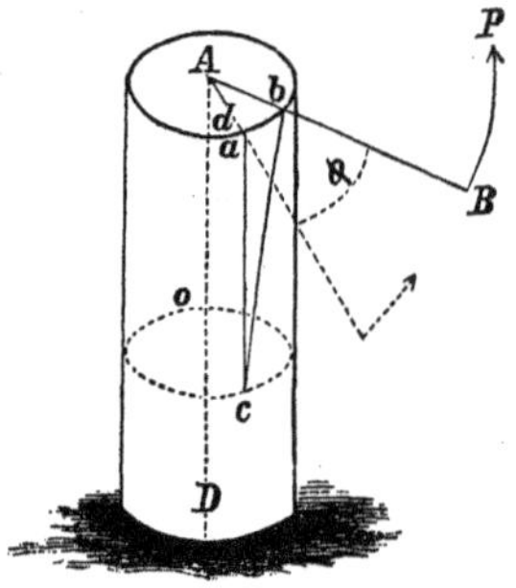

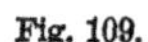

Fig. 109.

$f(\rho, \varphi) =$ the equation of a transverse section, and

G $=$ the coefficient of the elastic resistance to torsion, which is the force necessary to turn one end of a unit of area and unit of length of fibres through an angle unity, the vertex of the angle being on the axis of torsion, one end of the fibres being fixed and the twisting force being applied directly to the other end, and acting in the direction of a tangent to the arc of the path described by the free end.

As a unit of fibres cannot be placed so that all of them will be at a unit's distance from the axis, we must suppose that the resistance of a very thin annulus, which is at a unit's distance, is proportional to that of a unit of section; or the resistance

* Résumé des Leçons, Navier, Paris, 1864, p. 276 and several other pages following.

of an element at a units' distance from the axis is G multiplied by its area; which expressed analytically is

$$G\rho d\rho d\phi,$$

and according to the first law

$G\rho^2 d\rho d\phi =$ the resistance of any fibre whose length is unity, to being twisted through an angle unity; and the moment of resistance $= G\rho^3 d\rho d\phi$ for an angle unity; and for any angle θ the moment is, according to the second law,

$$G\theta\rho^3 d\rho d\phi$$

and the total moment equals the moment of the applied force, or moments of the applied forces; hence

$$Pa = G\theta \iint \rho^3 d\rho d\phi = G\frac{\alpha}{l} I_p,$$

where I_p is the polar moment of inertia of the section.

$$\text{For circular sections } I_p = \int_0^r \int_0^{2\pi} \rho^3 d\rho d\phi = \tfrac{1}{2}\pi r^4 \quad (199)$$

$$\therefore G = \frac{2Pa}{\pi\theta r^4} = \frac{2Pal}{\pi\alpha r^4} \quad (200)$$

$$\text{or, } \theta = \frac{2Pa}{G\pi r^4}$$

161. THE VALUE OF THE COEFFICIENT G may be found from equation (200). M. Cauchy found analytically on the condition that the elasticity of the material was the same in all directions, that $G = \frac{2}{5}$ E.* M. Duleau found experimentally that G is less than $\frac{2}{5}$ E, and nearly equal $\frac{1}{3}$ E, † and M. Wertheim found $G = \frac{3}{8}$ E nearly. ‡ M. Duleau's experiments gave the following mean values for G: †

	Pounds.
Soft iron	8,533,680
Iron bars	9,480,917
English steel	8,533,680
Forged steel (*very fine*)	14,222,800
Cast iron	2,845,600

---

* Résumé des Leçons, Navier, Paris, 1856, p. 197.

† Résistance des Matériaux, Morin, p. 461.

‡ L'Engineer, 1858, p. 52.

	Pounds.
Copper	6,209,670
Bronze	1,516,150
Oak	568,912
Pine	615,472

Example.—If an iron shaft whose length is 5 feet, and diameter 2 inches, is twisted through an angle of 7 degrees by a force P = 5,000 lbs., acting on a lever, $a = 6$ inches, required G. The 7 degrees is first reduced to arc by multiplying it by $\frac{\pi}{180}$, which gives $\alpha = \frac{7\pi}{180}$, and Eq. (200) gives,

$$G = \frac{2 \times 5000 \times 6 \times 60 \times 180}{(3.1416)^2 \times 7} = 9,697,000 \text{ lbs.}$$

162. TORSION PENDULUM.—If a prism is suspended from its upper end, and supports an arm at its lower end, and two weights each equal $\frac{1}{2}$ W are fixed on the arm at equal distances from the prism, and the prism be twisted and then left free to move, the torsional force will cause an angular movement of the arm until the fibres are brought to their normal position, after which they will be carried forward into a new position by the inertia of the moving mass in the weights $\frac{1}{2}$ W until the torsional resistance of the prism brings them to rest, after which they will reverse their movement, and an oscillation will result. The conditions of the oscillation may easily be investigated if the prism is so small that its mass may be neglected.

For, equation (200) readily gives:

$$P = \frac{\pi G r^4}{2la^2} (a\alpha)$$

from which it appears that the torsional force P varies as the space $(a\alpha)$ over which it moves.

It is a principle of mechanics that the moving force varies directly as the product of the moving mass multiplied by the acceleration. Hence, if $x = (a\alpha)$, the variable space, and t = the variable time, and M = the mass moved, and observing that t and x are inverse functions of each other, and the above principle of mechanics gives the following equation: —

$$M \frac{d^2x}{dt^2} = -P = -\frac{\pi G r^4}{2la^2} x.$$

Multiplying both members by the dx, gives

$$\frac{W}{g} \frac{dxd^2x}{dt^2} = -\frac{\pi G r^4}{2la^2} xdx,$$

where W is the weight of the mass moved, and g is the acceleration due to gravity. The oscillations commence at the extremity of an arc whose length is s, at which point the velocity is zero. The integral of the last equation between the limits s and x is

$$\frac{dx^2}{dt^2} = \frac{\pi G g r^4}{2Wla^2}(s^2 - x^2).$$

A second integral gives

$$t = \sqrt{\frac{2Wla^2}{\pi G g r^4}} \left[\sin^{-1}\frac{x}{s}\right]_0^s = \sqrt{\frac{\pi Wla^2}{2Ggr^4}}$$

which is the time of half an oscillation. For a whole oscillation:

$$2t = T = \frac{a}{r^2}\sqrt{\frac{2\pi}{Gg}lW}.$$

This is essentially the theory of Coulomb's torsion pendulum. A torsion pendulum was used by Cavendish in 1778 to determine the density of the earth. (See *Royal Philosophical Transactions:* London, Vol. 18, p. 388.) He found the mean density of the earth by this method to be 5.48 times that of water. This is considered the most reliable of all the known methods, but the *results* of other methods exceed the value given above by a small amount only, thereby confirming this result and showing that the mean density of the earth is about 5½ times that of water.

163. RUPTURE BY TORSION.—The resistance which a bar offers to a twisting force is a *torsional shearing resistance*, and in regard to rupture, the equation of equilibrium is founded upon the following principles:—

1st. The strain upon any fibre varies directly as its distance from the axis of torsion; and,

2d. The sum of the moments of resistances of the fibres equals the sum of the moments of the twisting forces.

Let S = the MODULUS OF TORSION, that is, the ultimate resistance to torsion of a unit of the transverse section which is most remote from the axis of torsion. It is the ultimate shearing resistance to torsion, but may be used for any shearing *strain* which is less than the ultimate,

d_1 = the distance of the most remote fibre from the axis of torsion,

$f(\rho, \phi)$ = the equation of the section,

P = the twisting force, and

a = the lever arm of P.

I_p = the polar moment of inertia of a section.

Then $\rho d\rho d\phi = dA$ = the area of an element of the section;

$S\rho d\rho d\phi =$ the shearing strain of the most remote element; and, by the first principle given above,

$\frac{S}{d_1}\rho d\rho d\phi =$ the shearing strain of any element, which is at a unit's distance from the axis of torsion, according to the first principle above; and from the same principle we have

$\frac{S}{d_1}\rho^2 d\rho d\phi =$ the shearing strain of any element, and this, multiplied by the distance of the element, ρ, from the axis, gives

$\frac{S}{d_1}\rho^3 d\rho d\phi =$ the moment of resistance to torsion.

Hence, according to the second principle we have

$$Pa = \frac{S}{d_1}\int\int\rho^3 d\rho d\phi = \frac{S}{d_1}\int\rho^2 dA. = \frac{S}{d_1} Ip \quad - \quad (201)$$

For circular sections, we have already found, Eq. (199),

$$Ip = \frac{1}{2}\pi r^4.$$

For square sections, whose sides are b, we may find *

$$Ip = \frac{1}{6}b^4, \text{ and } d_1 = b\sqrt{\tfrac{1}{2}}.$$

164. PRACTICAL FORMULAS.—Equations (199) and (201) give for cylindrical pieces, observing that $d_1 = r$,

$$Pa = \frac{1}{2}\pi S r^3 \therefore S = \frac{2\,Pa}{\pi r^3}. \quad - - - - - - - - \quad (202)$$

If cylindrical pieces are twisted off by forces which form a *couple*, and P, a, and r measured, the value of S may be found from equation (202). Cauchy found $S = \frac{4}{5}R$,† which is considered sufficiently exact when a proper coefficient of safety is used. Calling S = 25,000 pounds for iron, and using about a

---

* We have $\int\rho\ \ A = \int\left(x^2+y^2\right)dA = \int x^2dA + \int y^2dA$, that is, the polar moment equals the sum of the rectangular moments, the origin being the same in both cases. In this case the origin being at the centre of the square, we have $\int x^2dA = \int y^2dA \therefore Ip = 2\int y^2dA = 2\times\frac{1}{12}b^4$ (see Eq. (33)).

† Résumé des Leçons, Navier. Paris, 1856, pp. 193–203, and p. 507.

five-fold security; and S = 8,000 pounds for wood, and using about a ten-fold security, and we may use for

$$\left.\begin{array}{ll}\text{Round iron shafts (wrought or cast), diameter} & = \frac{1}{10}\sqrt[3]{Pa} \\ \text{Square iron shafts (wrought or cast), side of the square} & = \frac{1}{11}\sqrt[3]{Pa} \\ \text{Square wooden shafts, side of the squares} & = \frac{1}{8}\sqrt[3]{Pa}\end{array}\right\} \quad (203)$$

The dimensions given by these formulas are unnecessarily large for a steady strain, but shafts are frequently subjected to sudden strains, amounting sometimes to a shock, and in these cases the results are none too large.

Practical formulas may also be established on the condition that the *total angle of torsion* shall not exceed a certain amount. Making $G = \frac{3}{8}E$, and solving (200) in reference to r, and we have for cylindrical shafts,

$$r = \sqrt[4]{\frac{16\ Pal}{3\pi E\alpha}},$$

and similarly for square shafts,

$$b = \sqrt[4]{\frac{16\ Pal}{E\alpha}}$$

In these expressions P should not be so great as to impair the elasticity,—say for a steady strain P should not exceed the values given by equation (203).

If α° is given in degrees, it is reduced to arc by multiplying it by $\frac{\pi}{180}$ so that $\alpha = \frac{\pi}{180}\alpha^\circ$; hence the preceding equations become: for cylindrical iron shafts,

$$r = 3.14\sqrt[4]{\frac{Pal}{E\alpha^\circ}}; \quad (204)$$

and for square iron shafts,

$$d = 5.51\sqrt[4]{\frac{Pal}{E\alpha^\circ}}. \quad (205)$$

Examples—1. A round iron shaft fifteen feet long, is acted upon by a weight P = 2,000 lbs. applied at the circumference of a wheel which is on the shaft, the diameter of the wheel being two feet; what must be the diameter of the shaft so that the total angle of torsion shall be 2 degrees?

If the shaft is cast-iron E = 16,000,000, and

$$2r = d = 6.28 \sqrt[4]{\frac{2000 \times 12 \times 15 \times 12}{2 \times 16,000,000}} = 3.69 \text{ inches.}$$

2. A round wooden shaft, whose length is 8 feet, is attached to a wheel whose diameter is 8 feet. A force of 200 pounds is applied at the circumference of the wheel, what must be the diameter of the shaft so that the total angle of torsion shall not exceed 2 degrees?

$$2r = d = 6.28 \sqrt[4]{\frac{200 \times 4 \times 12 \times 8 \times 12}{2 \times 2,000,000}} = 4.35 \text{ inches.}$$

For further information upon this subject see "Résistance des Matériaux," Navier; Paris, 1856, pp. 237–509, and the exhaustive articles of Chevandier and Wertheim in "Annales des Chemie et Physique," Vol. XL. and Vol. L.

165. RESULTS OF WERTHEIM'S EXPERIMENTS.—A few years since M. G. Wertheim presented to the French *Académie des Sciences* an exhaustive paper upon the subject of torsion, the substance of which was published in the *Annales de Chimie et de Physique*, Vol. XXIII., 1st Series, and Vol. L., 3d Series. These articles would make a volume by themselves, and hence we will content ourselves at this time with presenting his

CONCLUSIONS.

When a body of three dimensions is subject to torsion the following facts are observed:—

1st. The torsion angle will consist of two parts, one temporary, the other permanent; the latter augments continually, though not regularly.

2d. The temporary displacements augment more and more rapidly than the moments of the applied couples, and the increase of the mean angle, which in hard bodies continues until rupture, in soft bodies continues only to the point where the body commences to suffer rapid and continuous deformation.

3d. The temporary angles are not rigorously proportional to the length, and, all else being equal, the disproportionality increases in measure as the bar becomes shorter.

4th. In all homogeneous bodies, torsion caused a *diminution of the volume*, which is proportional to the length and square of the angle of torsion, and each point of the body, instead of describing an arc of a circle, follows the arc of a spiral. The condensation of the body increases from the centre to the circumference.

5th. In bodies with three angles of elasticity, the change of volume and resistance to torsion are functions of the three axes, and the relation between them may be such that the volume will augment.

6th. Circular or turning vibrations of great amplitude are difficult to produce, and as small angles of torsion only are used, the preceding conclusions apply to this case.

7th. Rupture produced by torsion usually takes place at the middle of the length of the prism; it commences at the dangerous points, and operates by slipping in hard bodies and by elongation in soft ones.

8th. With regard to the influence of the figure and absolute dimensions of the transverse sections of the bodies, we derive the following conclusions:—

9th. In homogeneous circular cylinders the diminution of the volume is equal to the original volume multiplied by the product of the square of the radius, and the angle of torsion for a unit of length (the angle being always very small). Further, under torsion the radius of the cylinder equals the primitive radius multiplied by the sine of the angle of inclination of the helicoidal fibres. This last gives a means of calculating the diminution of volume. But in reality the twisted cylinder takes the form of two frustra of cones joined at the smaller bases; and although this does not sensibly affect the theoretical results for long cylinders, yet it deprives our formulas of all their value in ordinary practical cases.

CHAPTER IX.

EFFECT OF LONG-CONTINUED STRAINS—OF OFT-REPEATED STRAINS, AND OF SHOCKS—REMARKS UPON THE CRYSTALLIZATION OF IRON.

EFFECT OF LONG-CONTINUED STRAINS.

166. GENERAL EFFECT.—The values of the coefficients of elasticity and the modulii of tenacity, crushing, and of rupture were determined from strains which were continued for a short time—generally only a few minutes—or until equilibrium was apparently established; and yet it is well known that if the strain is severe, the distorsion, whether for extension, compression, or bending, will increase for a long time; and as for rupture, it always takes time to break a piece, however suddenly rupture may be produced. By sudden rupture we only mean that it is produced in a very short time.

The *increased* elongation due to a prolonged duration of the strain beyond a few minutes, will affect the coefficient of elasticity but very slightly, for the strains which are used in determining it are always comparatively small, and the greater part of the effect is produced immediately after the stress is applied. Still, if the distortion should go on indefinitely, no matter how slowly, the elasticity, and hence the coefficient, would be greatly modified by a very great duration of the stress, however small the stress may be; and at last rupture would take place. If the basis of this reasoning be well founded, we might reasonably fear the ultimate stability of all structures, and especially those in which there are members subjected to tension. But the continued stability of structures which have stood for centuries, teaches us, *practically at least*, that in all cases in which the strain is not too severe, equilibrium is established in a short time between the stresses and strains, and in such cases the piece will sustain the stress for an indefinitely long time.

167. **HODGKINSON'S EXPERIMENTS.**—The results of the experiments which are recorded in Article XL., page 48, show that in one case the compression increased with the duration of the strain for three-fourths of an hour. In the case of extension on another bar, as shown in Article VII., page 7, it appears that the same weight produced an increased elongation for nine hours; but during the last, or tenth hour, there was no increase over that at the end of the ninth hour.

In both these cases the strain was more than one-half that of the ultimate strength.

168. **VICAT'S EXPERIMENTS.**—M. Vicat took wrought-iron wire and subjected it to an uniform stress for thirty-three months. The elongations produced by the several weights were measured soon after the weights were applied, and total lengths determined from time to time during the thirty-three months. It was found for all but the first wire, as given in the following table, that the increased elongations after the first one were very nearly proportional to the duration of the stress. (*Annales de Chemie et Physique, Vol.* 54, 2*d series.*)

TABLE

Of the Results of M. Vicat's Experiments on Wrought-iron Wire.

Amount of Strain.	Very soon after the weights were laid on, the elongation of each piece was determined.	Increased Elongation after 33 months.
$\frac{1}{4}$ of its ultimate tensile strain...		No additional increase.
$\frac{1}{3}$ of its ultimate tensile strain...		0.027 of an inch per foot.
$\frac{1}{2}$ of its ultimate tensile strain...		0.040 of an inch per foot.
$\frac{3}{4}$ of its ultimate tensile strain...		0.061 of an inch per foot.

169. **FAIRBAIRN'S EXPERIMENTS.**—Fairbairn made experiments upon several bars of iron, which were subjected to a transverse strain, the results of some of which are recorded in the following tables. (*See Cast and Wrought-iron, by Wm. Fairbairn.*) The bars were four feet six inches between the supports, and weights were applied at the middle, and permitted to remain there several years, as indicated by the tables. The deflections were noted from time to time, and the results were recorded.

TABLE I.

In which the Weight Applied was 336 pounds.

TEMPERATURE.	Date of Observation.	Cold-blast,—deflection in inches.	Hot-blast,—deflection in inches.	Ratio of load to mean breaking weight.
	March 11, 1837....	1.270	1.461	Cold-blast,
78°	June 3, 1838......	1.316	1.538	0.661 : 1
72°	July 5, 1839.......	1.305	1.533	Hot-blast
61°	June 6, 1840......	1.303	1.520	0.694 : 1
50°	November 22, 1841.	1.306	1.620	
58°	April 19, 1842.....	1.308	1.620	
	Mean...........	1.301	1.548	

Previous to taking the observations in November and April the hot-blast bar had been disturbed.

In regard to this experiment Mr. Fairbairn remarks:—" The above experiments show a progressive increase in the deflections of the cold-blast bar during a period of five years of 0.031 of an inch, and of 0.087 of the hot-blast bar." The numerical results are found by comparing the first deflection with the mean of all the observed deflections. But an examination of the table shows that the greatest deflection, which was observed in both cases, was at the second observation, which was about a year and a quarter after the weight was applied, and during the next two years the *deflections decreased* 0.015 of an inch for the cold-blast, and 0.018 of an inch for the hot-blast bar. After this the deflections appear to increase for the cold-blast bar 0.005 of an inch the next two years. Considering all the particulars of these experiments it does not seem safe to conclude that the deflections would go on increasing indefinitely with a continuance of the load. Admitting that the small in crease of deflections during the last two years are correct and not due to errors of observation, and we see no reason why the deflections would not be as likely to decrease after a time as they were after the first year.

TABLE II.

In which the Bar was Loaded with 392 pounds.

TEMPERATURE.	Date of Observation.	Cold-blast,—deflection in inches.	Hot-blast,—deflection in inches.	Ratio of load to mean breaking weight.
	March 6, 1837.....	1.684	1.715	
78°	June 23, 1838.....	1.824	1.803	For cold-blast, 0.771 : 1
72°	July 5, 1839.......	1.824	1.798	
61°	June 6, 1840.......	1.825	1.798	For hot-blast, 0.805 : 1
50°	November 22, 1841.	1.829	1.804	
58°	April 19, 1842.....	1.828	1.812	
	Mean............	1.802	1.788	

Here we see a general increase in the deflections from year to year, being very regular in the cold blast, and quite irregular in the hot blast. But we observe that the increase is exceedingly small after the first year, being only 0.004 of an inch in the cold blast bar, and 0.009 of an inch in the other.

TABLE III.

TEMPERATURE.	Date of Observation.	Cold-blast,—deflection in inches.	Hot-blast,—deflection in inches.	Ratio of permanent load to mean breaking weight.
	March 6, 1837.....	1.410	Broke immediately with 448 pounds.	Cold-blast.
78°	June 23, 1838......	1.457		
72°	July 5, 1839.......	1.446		0.881 : 1
61°	June 6, 1840.	1.445		
50°	November 22, 1841.	1.449		
58°	April 19, 1842.....	1.449		
	Mean..........	1.442		

We find from this table, as from Table I., that the maximum deflection was observed about a year and a quarter after the weight was applied, and that it decreased during the next two years, after which it slightly increased. The deflections were the same at the two last observations. These changes took place under the severe strain of more than four-fifths of the breaking weight. These experiments *indicate* that for a steady strain which is less than three-fourths of the ultimate strength of the bar, the deflection will not increase progressively until rupture takes place, but will be confined within small limits.

170. ROEBLING'S OBSERVATIONS.—The old Monongahela bridge in Pennsylvania, after thirty years of severe service, was removed to make place for a new structure. The iron which was taken from the old structure was carefully examined and tested by Mr. Roebling, and found to be in such good condition that it was introduced by him into the new bridge.*

He also found that the iron in another bridge over the Alleghany river was in good condition after forty-one years of service.

* Roebling's Report on the Niagara Railroad Bridge, 1860, p. 17; Jour. Frank. Inst., 1860, Vol. LXX., p. 361.

OFT-REPEATED STRAINS.

Nearly all kinds of structures are subjected to greater strains at certain times than at others, and some structures, as bridges and certain machines, are subject to almost constant changes in the strains. Loads are put on and removed, and the operation constantly repeated. The only experiments to which we can refer for determining the effect of a load which is placed upon a bar and then removed, and the operation of which was frequently repeated, are those of Wm. Fairbairn, made in 1860.* The beam was supported at its ends, and the weight which produced the strain was raised and lowered by means of a crank and pitman, as shown in Fig. 110.

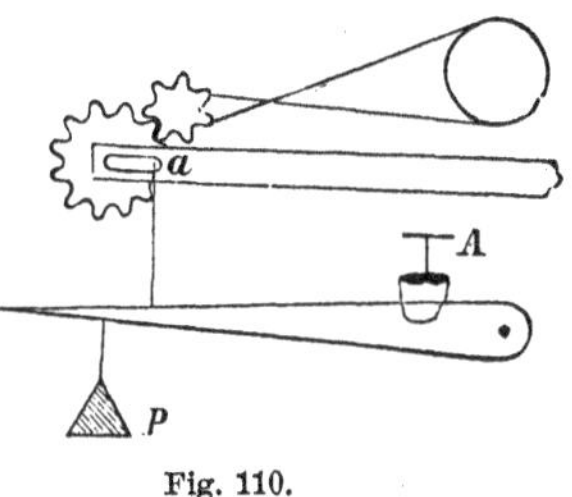

Fig. 110.

The gearing was connected with a water-wheel, which was kept in motion day and night, and the number of changes of the load were registered by an automatic counter. The beam was 20 feet clear span and 16 inches deep. The dimensions of the cross section were as follows:—

Top—Plate, $4 \times \frac{1}{2}=$	2.00	sq. inches
Angle irons, $2 \times 2 \times \frac{5}{16}=$	2.30	" "
Bottom—Plate, $4 \times \frac{1}{4}=$	2.00	" "
Angle Irons, $2 \times 2 \times \frac{3}{16}=$	1.40	" "
Web—Plate, $15\frac{1}{4} \times \frac{1}{8}=$	1.90	" "
Total	8.60	" "

Weight of beam, 1 cwt. 3 qr. 3 lbs.
Probable breaking weight, 9.6 tons.

First Experiment.—Beam loaded to $\frac{1}{4}$ the breaking weight:—

Total applied load	5,809 lbs.
Half the weight of the beam	434 "
Strain on the bottom flange	4.3 tons per sq. inch.
Margin of strength by Board of Trade	3.4

* Civ. Eng. and Arch. Jour., Vol. XXIII., p. 257, and Vol. XXIV., p. 237.

TABLE

Of the Results of Experiments made upon a Beam which was Supported at its Ends, and a Weight repeatedly but gradually Applied at the Middle.

DATE.	No. of Changes.	Deflection at Centre of Beam.	DATE.	No. of Changes.	Deflection at Centre of Beam.
1860.			1860.		
March 21..........	......	0.17	April 13..........	268,328	0.17
22..........	10,540	0.18	14..........	281,210	0.17
23..........	15,610	0.16	17..........	321,015	0.17
24..........	27,840	....	20..........	343,880	0.17
26..........	46,100	0.16	25..........	390,430	0.16
27..........	57,790	0.17	27..........	408,264	0.16
28..........	72,440	0.17	28..........	417,940	0.16
29..........	85,960	0.17	May 1..........	449,280	0.16
30..........	97,420	0.17	3..........	468,600	0.16
31..........	112,810	0.17	5..........	489,769	0.16
April 2..........	144,350	0.16	7..........	512,181	0.16
4..........	165,710	0.18	9..........	536,355	0.16
7..........	202,890	0.17	11..........	560,529	0.16
10..........	235,811	0.17	14..........	596,790	0.16

At this point, after half a million of changes, the beam did not appear to be damaged. At first it took a permanent set of 0.01 of an inch, which did not appear to increase afterwards, and the mean deflection for the last changes was less than for the first. For the last seventeen days the deflection was uniform, but for the first seventeen days it was variable.

The moving load was now increased to one-third the breaking weight, = 7,406 lbs., with the following results :—

DATE.	No. of Changes of Load.	Deflection in Inchs.	DATE.	No. of Changes of Load.	Deflection in Inchs.
1860.			1860.		
May 14..........	......	0.22	June 7..........	217,300	
15..........	12,623	0.22	9..........	236,460	0.21
17..........	36,417	0.22	12..........	264,220	0.21
19..........	53,770	0.21	16..........	292,600	0.21
22..........	85,820	0.22	21..........	327,000	0.22
26..........	128,300	0.22	23..........	350,000	0.23
29..........	161,500	0.22	25..........	375,650	0.25
31..........	177,000	0.22	26..........	403,210	0.23
June 4..........	194,500	0.21			0.23

The beam had now received 1,000,000 changes of the load, but it remained uninjured. The moving load was now increased to 10,050 lbs.—or one-half the breaking weight—and it broke with 5,175 changes. The beam was then repaired by riveting a piece on the lower flange, so that the sectional area was the same as before, and the experiment was continued. One hundred and fifty-eight changes were made with a load equal to one-half the breaking weight; and the load was then reduced to two-fifths the breaking weight, and 25,900 changes made. Lastly, the load was reduced to one-third the breaking weight, with the following results:—

DATE.	No. of Changes of Load.	Deflection in Inchs.	DATE.	No. of changes of Load.	Deflection in Inchs.
1860.			1860.		
August 13	25,900	0.18	Dec. 22	929,470	0.18
16	46,326	....	29	1,024,500	....
20	71,000	....	1861.		
24	101,760	....	Jan. 9	1,121,100	....
25	107,000	....	19	1,278,000	....
31	135,260	....	26	1,342,800	....
Sept. 1	140,500	....	Feb. 2	1,426,000	....
8	189,500	....	11	1,485,000	....
15	242,860	....	16	1,543,000	....
22	277,000	....	23	1,602,000	....
30	320,000	....	March 2	1,661,000	....
October 6	375,000	....	9	1,720,000	0.18
13	429,000	....	13	1,779,000	0 17
20	484,000	....	23	1,829,000	....
27	538,000	....	30	1,885,000	....
November 3	577,800	....	April 6	1,945,000	....
10	617,800	....	13	2,000,000	....
17	657,500	....	20	2.059,000	....
23	712,300	....	27	2,110,000	....
December 1	768,100	....	May 4	2,165,000	....
8	821,970	....	11	2,250,000	....
15	875,000	0.18	June	2,727,754	0.17

The piece had now received nearly 4,000,000 changes in all, but the 2,727,000 changes after it was once broken and repaired did not injure it. The changes were not very rapid. During the first experiment they averaged about 11,000 per day, or less than eight per minute, and during the last experiment the highest rate of change appears to have been less than eleven per minute, which is very slow compared with the stroke of some machines. Tilting-hammers often run from ten to twenty times this speed.

SHOCK—CRYSTALLIZATION.

171. SHOCKS.—When a weight is applied to another body *suddenly* it produces a " shock " upon the materials which compose the bodies. We cannot, practically, tell how frequently or with what force bodies must come in contact with each other in order to produce, a " shock ; " but theoretically any small body which is *suddenly* arrested in its movement, or *suddenly* deviated in its course by another body, produces a shock. Mass is necessary to the production of a shock, and the masses must impinge upon each other. If a *force* could manifest itself independent of matter, arresting the movement of it, however suddenly, by *another force*, would not produce a shock. Also, changing the movement of a mass by such a *force*, however sudden, will not produce a shock. These ideas are approximately realized in the movements of steam, air, and other gases. Steam impinges against air without producing shock, practically speaking. A moving piston (in some machines) is brought to rest by the reaction of steam, or by a steam cushion, without producing shock. The alternate expansions and contractions of a piston-rod, or pitman, or other similar piece in steam machinery, which are caused by the alternate pull and push of the moving force, do not produce a shock. The pieces may be "shocked" on account of working with loose connections, but that cause is not here considered. In the first example above cited there is very little mass in the moving or resisting bodies; in the next one the motion of the moving mass is changed, and may be brought quite suddenly to rest by the action of a highly elastic medium which has but little mass; and in the last example the particles are contiguous and are only slightly moved in reference to each other, as the forces of extension and compression are transmitted through the bar. The particles are not permanently displaced in reference to each other, as they are liable to be by a blow or "shock."

Shocks are practically prevented in many cases by the

introduction of elastic substances which possess considerable mass. Thus steel, rubber, and wooden springs in vehicles and in certain machines are familiar examples. But elasticity alone is not a sufficient protection, for, as has been previously observed, all bodies are elastic. When masses are used for springs they must be so arranged as to operate through a perceptible space in bringing the moving body to rest, or in changing its velocity a perceptible amount.

Springs, however elastic, will not always prevent a shock, although they may greatly relieve it. Thus, the springs under a car will not prevent the shock which always follows when the car-wheel *strikes* the end of a rail, although the shock is not as severe as it would be if the body of the car were rigidly connected with the axle. So, too, the springs between the buffers on a car and the body of the car will not prevent one buffer striking another so as to produce a "shock." In these cases the springs may prevent the shock from being transmitted in a large degree to the body of the car. The springs in certain forge-hammers operate in a similar way to prevent a "shock" upon the working parts of the machinery.

"Shocks" are very injurious to machinery, and hence should, so far as possible, be avoided. All machines in which "shocks" are necessary, or incidental, or accidental, such as steam forge-hammers, morticing machines, stone-drilling machines, and the like, are much more liable to break than those that operate by a steady pull and push. Metals are so liable to break under such circumstances that many have supposed that the internal structure is changed, and the metal becomes more or less crystallized.

The strength of the metal which is subjected to shocks is also greatly modified by the temperature—the lower the temperature the more damaging is the shock. It has been shown in Article 29, that wrought iron is somewhat stronger at a low temperature under a steady strain than at a higher temperature. Notwithstanding this is contrary to the "popular notion," it has been further confirmed by the very careful experiments of the "Committee appointed by His Majesty the King of Sweden," and reported by Knut Styffe, Director of the Royal Technological Institute of Stockholm.* Their first conclusion was:—† "The absolute strength (tenacity) of iron and steel is not diminished by cold, but that even at the lowest temperature which ever occurs in Sweden, it is at least as great as at the ordinary temperature—or 60° Fahr."

These results are confirmed by the more recent experiments of Joule of England, and by several other experimenters.

But it is generally supposed that machinery, railroad iron, tyres on locomotives, and tools, break much more frequently with the same usage when very cold than they do when warm. Is this a mere notion?

Is it because breakages are more annoying in cold than in warm weather, and hence make a more lasting impression upon the minds of those who have to deal with them; so that they think they occur more frequently? Impressions are not safe guides in scientific investigations. Our observations on the use of out-door machinery in cold and warm weather lead us to believe that they do break much more frequently with the same usage in winter than in summer. The same fact, in regard to the breaking of rails on railroads, was admitted by Styffe; but after arriving at the conclusion which he did in regard to the effect of cold upon the absolute strength of iron, he concluded that the cause of the more frequent breakages was due to the more rigid and non-elastic foundation caused by the frozen ground. But Sandberg, the translator of Styffe's work, thought that iron when subjected to shocks might not give the same relative strength at different temperatures that it would when subjected to a steady strain. He therefore instituted another series of experiments to satisfy himself upon this important point, and aid in solving the problem. The following is an abstract of his report:—

The supports for the rails in the experiments were two large granite blocks which rested upon granite rocks in their native bed. The rails were supported near their ends on these blocks. They were broken by a ball which weighed 9 cwt., which was permitted to fall five feet the first blow, and the height increased

---

* The Elasticity, Extensibility, and Tensile Strength of Iron and Steel. By Knut Styffe. Translated by Christer P. Sandberg, London.

† Ibid., p. 111.

one foot at each succeeding fall, and the deflection measured after each impact. A small piece of wrought iron was placed on the top of each rail to receive the blow, so as to concentrate its effect.

The rail was thus broken into two halves, and each part was afterwards broken at different temperatures. As the experiments were not made till the latter part of the winter, the lowest temperature secured was only 10° Fahr. Fourteen rails were tested:—Seven of which were from Wales; five from France; and two from Belgium. From these the experimenter drew the following conclusions:—*

1. "That for such iron as is usually employed for rails in the three principal rail-making countries (Wales, France, and Belgium), the breaking strain, as *tested by sudden blows or shocks*, is considerably influenced by cold; such iron exhibiting at 10° F., only one-third to one-fourth of the strength which it possesses at 84° F.

2. "That the ductility and flexibility of such iron is also much affected by cold, rails broken at 10° F. showing on an average a permanent deflection of less than one inch, whilst the other halves of the same rails, broken at 84° F., showed less than four inches before fracture."

This seems to be the fairest and most conclusive experiment upon this point that we have met with, but it is not satisfactory to all, or else they are ignorant of the experiment, for there has been of late considerable discussion upon the subject in the scientific journals. Some take the experiments of Fairbairn and Joule as conclusive upon the point, and attribute the cause of the failures, in many cases, to an inferiority of the iron, and in the case of tyres to an over-stretching of the metal when it is put on the wheel. By many, the presence of phosporous is considered especially detrimental to iron which is subjected to shocks in cold weather. But until the *fact* is established that cold iron is weaker than warm iron, when subjected to shocks, it is worse than useless to speculate upon the cause. More experiments are needed on this point, in which the quality of the metal, and all the conditions of the experiments should be definitely known.

---

* Styffe's work on iron and steel, translated by Sandberg, p. 157.

The following experiments, by John A. Roebling,* bear upon this subject, although they are not conclusive, for it is not reported that the same metal was tested when warm:—

"The samples tested were about one foot long, and were reduced at the centre to exactly three-fourths of an inch square, and their ends left larger, were welded to heavy eyes, making in all a bar three feet long. These were covered with snow and ice, and left exposed several days and nights. Early in the morning, before the air grew warmer, a sample inclosed in ice, was put into the testing-machine and at once subjected to a strain of 26,000 pounds, the bar being in a vertical position, and left free all around. The iron was capable of resisting 70,000 lbs. to 80,000 lbs. per square inch. A stout mill-hand struck the reduced section of the piece, horizontally, as hard as he could, with a billet one and a half inches in diameter and two feet long. The samples resisted from three to one hundred and twenty blows. With a tension of 20,000 lbs. some good samples resisted *300* blows before breaking."

The finest and best qualities of iron, or those that have the highest coefficient elasticity will resist vibration best. It is generally supposed that good iron will resist concussions much better than steel. Sir William Armstrong, of England, says:—

"The conclusion at which I have long since arrived, and which I still maintain, is, that although steel has much greater tensile strength than wrought iron, it is not as well adapted to resist concussive strains. It is impossible, then, that the vibratory action attending concussion is more dangerous to iron than to steel. The want of uniformity is another serious objection to the general use of steel in such cases." This has been used as an argument against the use of steel rails, but practically this has proved not to be a serious difficulty. So many elements must be considered in the use of steel rails, aside from fracture, that the problem must be solved by itself, and principally upon other grounds.

172. CRYSTALLIZATION.—A crystal is a homogeneous inorganic solid, bounded by plane surfaces, systematically arranged. The quartz crystal is a familiar example. Different substances

---

* *Jour. Frank. Inst.*, vol. xl., 3d series, p. 361.

crystallize in forms which are peculiar to themselves. Metals, under certain circumstances, crystallize; and if they are broken when in this condition the fracture shows small plane surfaces, which are the faces of the crystals. It is found in all cases that *crystallized iron is weaker than the same metal in its ordinary state.* By its ordinary state we mean that wrought iron is fibrous, and cast iron and steel are granular in their appearance.

Iron crystallizes in the cubical system.* Whöler, in breaking cast-iron plates readily obtained cubes, when the iron had long been exposed to a white heat in the brickwork of an iron smelting furnace.

Augustine found cubes in the fractured surface of gun barrels which had long been in use.

Percy found on the surface and interior of a bar of iron, which had been exposed for a considerable time in a pot of glass-making furnace, large skeleton octahedra. (He seems to differ from the preceding in regard to the form of the crystals.)

Prof. Miller, of Cambridge, found Bessemer iron to consist of an aggregation of cubes.

Mallet says:—"The plans of crystallization group themselves perpendicular to the external surfaces."

Bar iron will become crystalline if it is exposed for a long time to a heat considerably below fusion. Hence we see why large masses which are to be forged may become crystalline, on account of the long time it takes to heat the mass. Forging does not destroy the crystals, and forging iron at too low a temperature makes it tender, while steel at too high a temperature is brittle. The presence of phosphorous facilitates crystallization. Time, in the process of breaking iron, will often determine the character of the fracture. If the fracture is slow, the iron will generally appear fibrous; but if it be quick, it will appear more or less crystalline. Many mechanics have noticed this result. At Shoeburyness armor-plates were shattered like glass under the impact of shot at a velocity of 1,200 feet to 1,600 feet per second. The iron was good fibrous iron.

Many engineers are of the opinion that oft-repeated and long-

---

* See Osborn's Metallurgy, pp. 83–86.

continued shocks will change fibrous to crystalline iron. Opinions, however, are divided upon this subject. In view of the immense amount of machinery and other constructions, parts of which are constantly subjected to shocks, the importance of the subject can hardly be overestimated.

William Fairbairn says:—* "We know that in some cases wrought iron subjected to continuous vibration assumes a crystalline structure, and that then the cohesive powers are much deteriorated; but we are ignorant of the causes of this change."

The late Robert Stephenson † referred to a beam of a Cornish engine which received a shock eight or ten times a minute, equal to about fifty tons, for a period of twenty years without apparent change. These shocks were not very frequent, and would not be considered as detrimental as if they occurred a few times each second. He also says:—" The connecting-rod of a certain locomotive engine that had run 50,000 miles, and received a violent jar eight times per second, or 25,000,000 vibrations, exhibited no alteration." In all the cases investigated by him of supposed change of texture, he knew of no single instance where the reasoning was not defective in some important link. These are not fair examples of shocks, as the vibrations referred to seem to be only changes from tension to compression, and the reverse.

Mr. Brunel accepted the theory of molecular change, for a time, as due to shocks, but afterwards expressed great doubts as to its correctness, and thought that the appearance depended more upon the manner of breaking the metal than upon any molecular change.

Fairbairn has speculated a little upon the probable cause of the internal change when it takes place. In *his evidence before the Commissioners appointed to inquire into the application of iron to railway structures*, he says:—" As regards iron it is evident that the application and abstraction of heat operates more powerfully in effecting these changes than probably any other agency; and I am inclined to think that we attribute too much influence to percussion and vibration, and neglect more obvious causes which are frequently in operation to produce

---

* *Civ. Eng. and Arch. Jour.*, vol. iii. p. 257.

† *Am. R. Times*, March 6, 1869, Boston.

the change. For example, if we take a bar of iron and heat it red hot, and then plunge it into water, it is at once converted into a crystallized instead of a fibrous body; and by repeating this process a few times, any description of malleable iron may be changed from a fibrous to a crystalline structure. Vibration, when produced by the blows of a hammer or similar causes, such as the percussive action upon railway axles, I am willing to admit is considerable; but I am not prepared to accede to the almost universal opinion that granulation is produced by those causes only. I am inclined to think that the injury done to the body is produced by the weight of the blow, and not by the vibration caused by it. If we beat a bar with a small hammer, little or no effect is produced; but the blows of a heavy one, which will shake the piece to the centre, will probably give the key to the cause which renders it *brittle*, but probably not that which causes *crystallization*. The fact is, in my opinion, we cannot change a body composed of a fibrous texture to that of a crystalline character by a mechanical process, except only in those cases where percussion is carried to the extent of producing considerable increase of temperature. We may, however, shorten the fibres by continual bending, and thus render the parts brittle, but certainly not change the parts which were originally fibrous into crystals.

For example, take the axle of a car or locomotive engine, which, when heavily loaded and moving with a high velocity, is severely shocked at every slight inequality of the rails. If, under these circumstances, the axle bends—however slightly—it is evident that if this bending be continued through many thousand changes, time only will determine when it will break. Could we, however, suppose the axle so infinitely rigid as to resist the effects of percussion, it would then follow that the internal structure of the iron will not be injured, nor could the assumed process of crystallization take place."

The late John A. Roebling, the designer and constructor of the Niagara Railway Suspension Bridge, in his report on that structure in 1860,* says he has given attention to this subject for years, and as the result of his observation, study, and experiment, gives as his view that "a molecular change, or so-called

* *Jour. Frank. Inst.*, vol. xl., 3d series, p. 361.

granulation or *crystallization*, in consequence of vibration or tension, or both combined, has in no instance been satisfactorily proved or demonstrated by experiment." "I further insist that crystallization in iron or any other metal *can never take place in a cold state.* To form crystals at all, the metal must be in a highly heated or nearly molten state." Notwithstanding these positive statements, he still hesitates to express a decided opinion which will cover the whole field of investigation. Still further on he states that he is witnessing the fact daily that vibration and tension combined will greatly affect the strength of iron *without changing its fibrous texture.* Wire ropes and iron bars will become weakened as the *vibration and tension* to which they are subjected increase.

Certain machines in which the working parts are subjected to frequent shocks, more or less severe, are constantly failing, and the general impression is that the failure is due to crystallization. In speaking of the rock-drilling engines used in Hoosac Tunnel, Mass., which were driven by compressed air, the committee say: *—" Gradually they begin to fail in strength; the incessant and rapid blows—counted by millions—to which they are subjected, appearing to *granulate* or *disintegrate* portions of the metals composing them." Having had some experience with this class of machines, I know something of the difficulties which surround them.† During the winter of 1866–7 my assistant in the University, Professor S. W. Robinson, and myself made several experimental machines, in the use of which we learned many essential conditions which must be observed in order to avoid frequent breakages in the use of iron which is subjected to frequent and long-continued shocks. As first designed, the breakages in the several working parts were exceedingly numerous, the remedy for which was not in making those parts larger and stronger, for that only aggravated the evil in most cases, but in arranging the moving parts so that they would be moved and brought to rest with as little shock as possible, and then making them as light as possible consistent

---

* Annual Report of the Commissioners on the Troy and Greenfield Railroad and Hoosac Tunnel. House Doc., No. 30, p. 5, Boston, Mass.

† Notice of, by B. H. Latrobe, C. E. Sen. Doc. No. 20, 1868, p. 31, Boston, Mass.

with strength. But at the rapid rate at which we ran them, which was from 300 to over 500 blows per minute, it was not easy to comply with these conditions. No especial difficulty was experienced from the general shock caused by the striking of the tool upon the rock; but the chief difficulty arose from the blow or shock to which the working pieces were directly subjected in operating them. At last, however, according to the report of the superintendent of the Marmora Iron Mine, Ont., at which place they have been in use for some time, we so far overcame all the difficulties as to make it a decidedly practical machine. In the experiments we found that it was a bad condition to subject a piece to a blow crosswise of its length; that is, perpendicular to its length.* It was also found that a piece struck obliquely would sustain a much greater number of blows than if struck perpendicularly. Many pieces evidently broke slowly, and was analogous to breaking a piece of tough iron on an anvil by comparatively light blows. If the blows were so severe as to start a crack in the piece, it would ultimately break if the blows were continued sufficiently long. Several of the broken pieces were critically examined to see if they were crystallized, but there were no indications of any change in the internal structure of the metal.

All sharp angles in pieces which are shocked should, if possible, be avoided; for in the process of manufacture they are liable to be rendered weaker at such points, and if they are equally strong so far as manufacture is concerned, a greater strain, at the instant they are shocked, is liable to fall at such points, thus rendering them relatively weaker there. At least it is found that such pieces are more liable to break at the angle. Hence, in the construction of direct-action rock-drills, direct-action steam-hammers, and similar percussion machines, the steam piston is not only generally made solid with the rod, but it is connected with it by a curve. In other words, the rod is more or less gradually enlarged into a piston. At first, much difficulty was experienced in this regard with steam-hammers,

* I have seen it published that a small hammer was made to strike a blow upon the side of a bar which was suspended vertically, the blows being repeated night and day for nearly a year, when the bar broke; but as the force of the blow and size of the bar were not given, I have thought that the statement was too indefinite to be of any scientific value.

for in some cases the piston broke away from the rod, and slipped down over it.

In certain steam forge-hammers the piston-rod is liable to break where it joins the hammer. In this case the worst possible arrangement is to make a rigid connection between the hammer and rod. The author once saw a rod three and one-half inches in diameter nicely fitted into a conical hole in a 400-pound hammer, the taper of the rod and hole being slight, so that it would hold by friction when once driven into place. The connection was practically rigid. The rod broke twice with ordinary use inside of twenty days. Probably it would have lasted a long time if the blow could always have been exactly central; but in ordinary use it would very naturally be subjected to cross strains by making a blow when the material was under one edge of the hammer. By a repetition of these cross strains, rupture might have been produced without any crystallization.

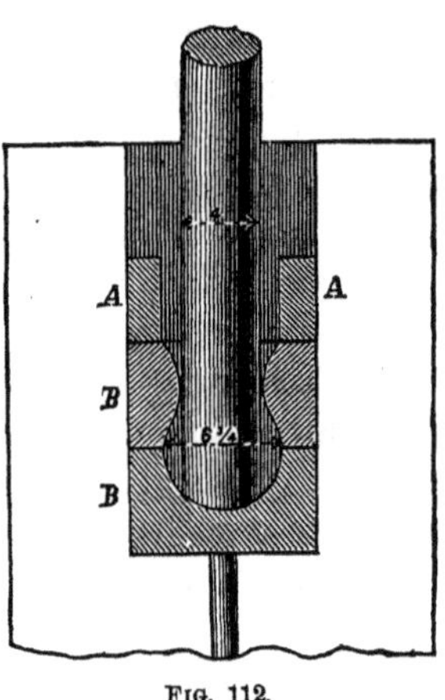

FIG. 112.

This difficulty is practically overcome in several ways, one of the most common of which is to place blocks of wood or other slightly elastic body between the end of the piston-rod and the hammer. Morrison overcame the difficulty by making the rod so large that it (the rod) became the hammer, and a small block of iron or steel was fitted into the forward end of the rod to serve as a face for the hammer.

A Mr. Webb, of England, proposed to overcome the difficulty by a device which is shown in Fig. 112.

Referring to the figure, it will be seen that the piston-rod, which is for the main part of its length 4 in. in diameter, is enlarged at the lower end to 6¾ in. in diameter, and is shaped spherically. This spherical portion of the rod is embraced by the annealed steel castings, B B, which are secured in their place in the hammer-head by the cotters, A, and the whole thus forms a kind of ball-and-socket joint, which permits the hammer head to swivel slightly on the rod without straining the latter. Mr. Webb first applied this form of hammer-rod fastening to a five-ton Nasmyth hammer with a 4 in. rod. With the old mode of attachment, with a cheese end, this hammer broke a rod every three or four weeks when working steel, while a rod with the ball-and-socket joint, which was put in in November, 1867, has been working ever since, that is, to some time in 1869, without giving any trouble. The inventor has also applied a rod thus fitted to a five-ton Thwaites and Carbutt's hammer with equal success.

These examples seem to indicate that if iron is crystallized on account of shocks, the progress of the change may be slackened by a judicious arrangement of the pieces and by proper connections. But it does not follow that because a metal breaks so frequently when subjected to shock, that it has become crystallized. It has been observed by those who have made the experiment, that a piece of bar iron which is broken by heavy blows, when the piece is so supported as to bound with each stroke, will present a crystalline fracture ; but if the same bar be broken by easy blows *near the place* of the former fracture, it will present a fibrous texture ; showing that in the former case the internal structure was not changed, unless it were in the immediate vicinity of the fracture. In such cases the *appearance of the surface of the fracture* does not indicate the true state of the internal structure. One reason why metals fail which are subjected to concussions, without crystallizing, is ;—an excessive strain is brought upon some point, thus impairing the elasticity and weakening the resisting powers, in which case, if the strains be repeated sufficiently long, rupture must ultimately take place. Or, if the concussion be sufficiently severe and local, it may displace the particles, and thus begin a fracture. This frequently takes place in the case of anvils, hammers, hammer-blocks, and the like.

If a bar is bent by a blow sufficiently great to produce a set, and the bar be bent back by another blow, and so on, the bar being bent alternately to and fro, rupture would probably take place at some time, however remote. It is often difficult to determine the strain which falls at a particular point of a piece when it is subjected to a shock, but if we could determine its exact amount, we might find it to be sufficiently large to account for rupture by shocks, without considering any mysterious change in the internal structure of the metal.

Note. Since the above was written, several articles have appeared in the scientific journals, giving the results of observations and experiments upon the strength of iron at low temperatures, and they all confirm the position above stated,—that iron will not resist shocks as effectually at very low temperatures as it will at ordinary temperatures.

CHAPTER X.

LIMITS OF SAFE LOADING OF MECHANICAL STRUCTURES.

173.—RISK AND SAFETY.—We have now considered the breaking-strength of materials under a variety of conditions, and also the changes produced upon them when the strains are within the elastic limits. In a mechanical structure, in which a single piece, or a combination of pieces, are required to sustain a load, it is desirable to know how small the piece, or the several pieces, may be made to sustain a given load *safely* for an indefinite time; or, how much a given combination will sustain *safely*. The nature of the problem is such that an exact limit cannot be fixed. Materials which closely resemble each other do not possess *exactly* the same strength or stiffness; and the conditions of the loading as to the amount or manner in which it is to be applied, may not be exactly complied with. Exactness, then, is not to be sought; but it is necessary to find a limit below which, in reference to the structure, or above which, in reference to the load, it is not safe to pass.

It is evident that to secure an economical use of the material on the one hand, and ample security against failure on the other, the limit should be as definitely determined as the nature of the problem will admit; but in any case we should incline to the side of safety. No doubt should be left as to the stability of the structure. *There is no economy in risk* in permanent structures. Risk should be taken only in temporary, or experimental, structures; or where risk cannot, from the nature of the case, be avoided.

174. ABSOLUTE MODULAS OF SAFETY.—In former times, one of the principal elements which was used for securing safety in a structure, was to assume some arbitrary value for the resistance of the material, such value being so small that the material could, in the opinion of the engineer, safely sustain it. This is a convenient mode, but very unphilosophical, although still extensively used. The plan was to determine, as nearly as possible, what good materials would sustain for a long period, and use that value for all similar materials. But it is evident, from what has been said in the preceding pages, that some materials will sustain a much larger load than the average, while others will not sustain nearly so much as the average. In all such cases the proper value of the *modulas* can only be determined by direct experiment. In all important structures the strength of the material, especially iron and steel, should be determined by direct experiment.

The following values are generally assumed for the *modulas of safety.*

		Pounds per square inch.
Wrought, iron *for tension or compression*, from		10,000 to 12,000
Cast iron, for *tension*, from		3,000 to 4,000
Cast iron, for *compression*, from		15,000 to 20,000
Wood, *tension or compression*, from		850 to 1,200
Stone, *compression*	granite, from	400 to 1,200
	quartz, from	1,200 to 2,000
	sandstone, from	300 to 600
	limestone, from	800 to 1,200

The practice of French engineers,* in the construction of bridges, is to allow 3.8 tons (gross) per inch upon the gross section, both for tension and compression of wrought iron.

The Commissioners on Railroad Structures, England, established the rule that the maximum tensile strain upon any part of a wrought iron bridge should not exceed five tons (gross) per square inch. †

In most cases the *effective section* is the section which is subjected to the strain considered.

175. FACTOR OF SAFETY.—The next mode, and one which is also largely in use, is to take a fractional part of the ultimate strength of the material, for the limit of safety. The reciprocal of this fraction is called *the factor of safety.* It is the ratio of the ultimate strength to the computed strain, and hence is the

* Am. R. R. Times, 1871, p. 6.

† Civ. Eng. and Arch. Jour. Vol. xxiv., p. 327.

factor by which the computed strain must be multiplied to equal the actual strength of the material, or of the structure.

Experiments and theory combine to teach that the *factor of safety* should not be taken as small as 2. See articles 19, 166, 167, and 168.

Beyond this the *factor* is somewhat arbitrarily assumed, depending upon the ideas of the engineer. For instance, the following values were given to the Commissioners on Railway Structures, in England.*

	Factors.
Messrs. May and Grissel	3
Mr. Brunell	3 to 5
Messrs. Rasbrick, Barlow and others	6
Mr. Hawkshaw	7
Mr. Glyn	10

The following values are also given by others :—

		Factors.
Bow, for wrought-iron beams		3.5
Weisbach, for wrought iron †		3 to 4
Vicat, for wire suspension bridges		more than 4
Rankine, for wire bridges	steady strain	3 to 4
	moving load	6 to 8
Fink, iron-truss bridges	for posts and braces	5 to 6
	for cast-iron chords	7
Fairbairn, for cast iron beams ‡		5 to 6
C. Shaler Smith, compression of cast iron		5
Rankine and others, for cast-iron beams		4 to 6
Mr. Clark in Quincy Bridge, lower chord		6 to 7
Washington A. Roebling, for suspension cables		6
Morin, Vicat, Weisbach, Rondelet, Navier, Barlow, and many others say that for a wooden frame it should not be less than		10
For stone, for compression		10 to 15

From the experiments which are recorded in Article 170, Fairbairn deduced the following conclusions in regard to beams

* Civ. Eng. and Arch. Jour., Vol. xxiv, p. 327.

† Weisbach, Mech. and Eng. Vol. 1, p. 201.

‡ Fairbairn, Cast and Wrought Iron, p. 58.

and girders, whether plain or tubular. * "The weight of the girder and its platform should not in any case exceed one-fourth the breaking weight, and that only one-sixth of the remaining three-fourths of the strength should be used by the moving load." According to this statement the maximum load, including the live and dead load, may equal, but should not exceed,

$$\tfrac{1}{4} + \tfrac{1}{6} \text{ of } \tfrac{3}{4} = \tfrac{3}{8}$$

of the breaking load. Hence the *factor of safety* must not be less than 2.66 when the above conditions are fulfilled. This value is, however, evidently smaller than is thought advisable by most engineers.

The rule adopted by the Board of Trade, England, for railroad bridges is † "to estimate the strain produced by the greatest weight which can possibly come upon a bridge throughout every part of the structure which should not exceed *one-fifth the ultimate strength of the metal.*" They also observed that ordinary road bridges should be proportionately stronger than ordinary railroad bridges.

176. RATIONAL LIMIT OF SAFETY.—It is evident that materials may be strained any amount within the elastic limit. Their recuperative power—if such a term may properly be used in connection with materials—lies in their elasticity. If that is damaged the life of the material is damaged, and its powers of resistance are weakened. As we have seen in the preceding pages, there is no known relation between the coëfficient of elasticity, and the ultimate strength of materials. The coëfficient of elasticity may be high and the modulus of strength comparatively low. In other words, the limit of elasticity of some metals may be passed by a strain of less than one-third their ultimate strength, while in others it may exceed one-half their ultimate strength. We see, then, the unphilosophical mode of fixing an *arbitrary modulus of safety*, or even a *factor of safety*, when they are made in reference to the ultimate strength. But an examination of the results of experiments shows that the limit of elasticity is rarely passed for strains which are less than one-third of the ultimate strength of the metal, and hence, according

* Civ. Eng. and Arch. Jour., Vol. xxiv., p. 329.

† Civ. Eng. and Arch. Jour., Vol. xxiv., p. 226.

to the views of the engineers given in the preceding article, the *factor of safety* is generally safe. But if the limit of elasticity were definitely known it is quite possible that a smaller *factor of safety* might sometimes be used.*

This method of determining the limit has been recognized by some writers, and the propriety of it has been admitted by many practical men, but the difficulty of determining the elastic limit has generally precluded its use. The experiments which are necessary for determining it are necessarily more delicate than those for determining the ultimate strength.

There is also a slight theoretical objection to its use. The limit of elasticity is not a definite quantity, for it is not possible to determine the *exact* point where the material is overstrained. But this is not a fatal objection, for the limit can be determined within small limits.

In regard to the margin that should be left for safety, much depends upon the character of the loading. If the load is simply a dead weight, the margin may be comparatively small; but if the structure is to be subjected to percussive forces or shocks, it is evident, as indicated in articles 19 and 171, that the margin should be comparatively large, not only on account of the indeterminate effect of the force, but also on account of the effect of such a force upon the resisting powers of the material. In the case of railroad bridges, for instance, the vertical posts or ties, as the case may be, are generally subjected to more sudden strains due to a passing load, than the upper and lower chords, and hence should be relatively stronger. The same remark applies to the inclined ties and braces which form the trussing; and to any parts which are subjected to severe local strains.

The frames of certain machines, and parts of the same machines, are subjected to a constant jar while in use, in which cases it is very difficult to determine the proper margin which is consistent with economy and safety. Indeed, in such cases, economy as well as safety generally consists in making them

---

* James B. Eads, in his *Report upon the Illinois and St. Louis Bridge*, for 1871, states that he tested samples of steel which were to be used in that structure, which showed limits of elastic reaction of 70,000 to 93,000 pounds per square inch.

excessively strong, as a single breakage might cost much more than the extra material necessary to fully insure safety.

The mechanical execution of a structure should be taken into consideration in determining the proper value of the margin of safety. If the joints are imperfectly made, excessive strains may fall upon certain points, and to insure safety, the margin should be larger. No workmanship is *perfect*, but the elasticity of materials is favorable to such imperfections as necessarily exist; for, when only a portion of the surface which is intended to resist a strain, is brought into action, that portion is extended or compressed, as the case may be, and thus brings into action a still larger surface. But workmanship which is so badly executed as to be considered *imperfect* would fail before all its parts could be brought into bearing.

176. EXAMPLES OF STRAINS THAT HAVE BEEN USED IN PRACTICAL CASES. The margin of safety that has been used in various structures may or may not serve as guides in designing new structures. If the margin for safety is so small that the structure appears to be insecure and gives indications of failure, it evidently should not be followed. It serves as a warning rather than as a guide. If the margin is evidently excessively large, demanding several times the amount of material that is necessary for stability, it is not a guide. Any engineer or mechanic, without regard to scientific skill or economy in the use of materials, may err in this direction to any extent. But if the margin appears reasonably safe, and the structure has remained stable for a long time, it serves as a valuable guide, and one which may safely be followed under similar circumstances. Structures of this kind are practical cases of the approximate values of the inferior limits of the *factors of safety*. The following are some practical examples:—

IRON TRUSSED BRIDGES.

NAME OF THE BRIDGE.	TENSION. Tons per square inch.	COMPRESSION. Tons per square inch.
Passaic (*Lattice*)	5½ to 6	4¼ to 5¼
Place de l'Europe (*Lattice*)	4	3⅜
Canastota (*N. Y. C. R. R.*) (*Lattice*)	5	4
Newark Dyke (*Warren Girder*)	5	5
Boyne Viaduct (*Lattice*)	5	
Charring Cross (*Lattice*)	5	4
	Pounds per square inch.	Pounds per square inch.
St. Charles, Mo. (*Whipple Truss*) *	12,000	12,000
Louisville, Ky. (*Fink Truss*)	7,000 to 12,000	⅕ to ⅐ the strength
Keokuk and Hannibal †	9,251	8,962
Quincy Bridge ‡	10,000	Factor of safety, 5
Kansas City Bridge §	11,375	711
Hannibal Bridge (*Quadrangular Truss*) ‖	Factor of safety, 5	Factor of safety, 5

WOODEN BRIDGES.

NAME OF THE BRIDGE.	MAXIMUM STRAIN.
Cumberland Valley R. R. Bridge	635 pounds per square inch.
Portage Bridge (*N. Y. & E. R. R.*)	Factor of safety 20.

---

* *R. R. Gazette*, July 8, 1871, p. 169.

† *R. R. Gazette*, July 15, 1871, p. 178. Pivot span 376 feet 5 inches; longest pivot span yet constructed.

‡ Report of Chief Engineer Clark.

§ Calculated from the Report of Chief Engineer O. Chanute, pp. 106 and 136.

‖ The tensile strength of the material ranged from 55,000 lbs. to 65,000 lbs. per square inch.—*R. R. Gazette*, July 15, 1871, p. 169.

CAST-IRON ARCHES.*

NAME OF THE ARCH.	SPAN. Feet.	SPAN. Inches.	VERSED SINE. Feet.	VERSED SINE. Inches.	STRAIN PER SQUARE INCH IN TONS.
Austerlitz	106	0	10	7	2.78
Carrousal	152	2	16	1	1.46
St. Denis	102	5	11	4	1.37
Nevers	137	9	15	0	1.90
Rhone	197	10	16	5	2.37
Westminster	120	0	20	0	3.00

STONE ARCHES.†

NAME OF THE ARCH.	Span in feet.	Versed sine in feet.	Pressure per square inch in pounds at the key.	Factor of safety at the point of greatest strain.
Wellington	100	15	175	11 3
Waterloo (9 *Arches*)	120	35	151	20.0
Neuilly	128	32	172	11.6
Taaf (*South Wales*)	140	35	244	8.0
Turin	147	18	293	10.2
London	152	38	215	14.0
Chester	200	42	349	8.6

CAST-STEEL ARCH.

NAME OF ARCH.	SPAN. Feet.	FACTOR OF SAFETY.
Illinois and St. Louis Bridge	515	6+‡

---

* Irwin on Iron Bridges and Roofs.

† Cresy's Encyclopædia.

‡ Report of the Engineer, p. 33.

SUSPENSION BRIDGES.

NAME OF THE BRIDGE.	Span in feet.	Strain in tons per square inch. From Bridge.	Strain in tons per square inch. Bridge and Load.	Factor of safety.
Menai	580	4.21	8.00	3.9*
Hammersmith	422½	5.38	9.36	3.3*
Pesth	666	5.01	8.11	3.9*
Chelsea	384	4.36	8.07	3.9*
Clifton	702¼	2.90	5.03	6.4*
Niagara	821	6.70	8.40	5.3†
Suspension Aqueduct, Pittsburgh, Pa. 7 spans, each,	160	....	....	4.0
Cincinnati Bridge‖	1,057	9.1	11.7	6 2
East River	1,600‡	....	....	6.0
Highland (*proposed*)	1,665§	....	....	6.0

TUBULAR BRIDGES.

NAME OF BRIDGE.	SPAN. Feet.	FOR WEIGHT OF BRIDGE AND LOAD.	
		Tension. Tons.	Compression. Tons.
Conway	400	6.85	5.03
Britannia (Central span)	460	3.00	....
Penrith (Tubular Girder)	...	4.75	4.25

---

* Tensile strength, 70,000 lbs. per square inch.
† Tensile strength, 100,000 lbs. per square inch.
‡ Engineer's Report. Suspending ties, factor of safety, 8.
§ Jour. Frank. Inst., vol LXXXVII., p. 165.
‖ Report of the Chief Engineer, J. A. Roebling.

STONE FOUNDATIONS.

	FACTOR OF SAFETY.
Pillars of the Dome of St. Peter's (*Rome*)...............	16
" " " St. Paul's (*London*).............	14
" " " St. Geneviève (*Paris*)............	7.6
Pillars of the Church Touissaint (*Angers*)...............	10
Merchants' Shot Tower (*Baltimore*)*..................	4.8
Lower courses of Britannia Bridge......................	31
Lower courses of the Piers of Neuilly Bridge (*Paris*).....	15.8
Foundation of St. Charles' Bridge (*Missouri*)...........	12 to 14
Foundations of East River Bridge†.....................	10 to 20

177. PROOF LOAD. The proof load is a trial load. It is intended as a practical test of a theoretical structure.

It generally exceeds the greatest load that it is ever intended to put upon the structure. Some structures, especially steam boilers, are subjected to excessively severe trial strains, often being two or even three times that to which they are to be subjected in actual use. "This is done," they say, "to insure against failure." But such severe strains do no good, and often do great damage, for they may overstrain the material and thus weaken it. For instance, if a structure is proportioned to carry 10,000 pounds per square inch, but on account of the imperfection of the material or imperfect workmanship, some point is required to carry 20,000 pounds per square inch; and if the structure is tested to 20,000 pounds per square inch, that point would be obliged to carry nearly 40,000 pounds per square inch, an amount which it might not be able to sustain for a long time. We say nearly 40,000 pounds; for it should be remembered that where the workmanship is imperfect, thereby throwing greater strains on certain points than was intended, the elasticity of the materials permits those members to be compressed or extended more than they otherwise would be, and thus tends

* Strength of materials, *J. K. Whildin*, p. 23.

† "In the stone work the pressures vary from 8 to 26 tons per square foot. Stone used is granite, selected samples of which have borne a crushing strain of 600 tons per square foot. Some will not bear over 100 tons per square foot. The general average is necessarily much less than that of the best specimens." —*Statement of the Chief Engineer, Washington A. Roebling.*

to bring into action the other members which at first were more lightly strained. In this way there is a *tendency* to bring about an equilibrium of strains on all those parts which were calculated to carry equal amounts.

According to the principles which have been discussed in the preceding pages, it is evidently better for the structure, and should be more satisfactory, to apply a moderate *proof load* for a long time than an *excessive* one for a short time.

APPENDIX I.

PRESERVATION OF TIMBER.

(The *Graduating Thesis* of Mr. H. W. Lewis, of the class of 1866, and more recently engineer on the Missouri Valley Railroad, forms the basis of this article. I have added to it such new matter as I find in the *Graduating Thesis* of Mr. A. B. Raymond, of the class of 1871.—AUTHOR.)

1. **THE IMPORTANCE OF THIS SUBJECT** may be shown by many familiar examples in practical life.

Although iron is coming more and more extensively into use, yet the amount of wood which is used at the present time in mechanical structures, and which will, in the nature of things, be used for a long time to come, is enormous. For instance: in 1865 there was sold in Chicago alone 900,000,000 pieces of lath, 2,000,000,000 of shingles, and 5,000,000,000 feet of lumber.*

In the matter of railroad ties alone, any process which could be easily and cheaply applied, which would double their life, would literally save millions to the country. This may be shown by an approximate calculation, thus:—Allowing only 2,000 sleepers to the mile, at a cost of fifty cents each, and admitting that the average life of American sleepers is only seven years,† and that it costs ten cents to treat each tie in some way so as to make it last fourteen years, then the saving at the end of seven years is $600 per mile. For ten cents at compound interest at ten per cent. for seven years amounts to twenty cents, which from fifty cents leaves thirty cents as the net saving on one; and on 2,000 it amounts to $600.

There are in the United States about 45,000 miles of railroad; and hence, if the above conditions could be realized of all of them, the annual saving would be about $3,400,000! Other uses of timber would show a corresponding saving.

2. **CLASSIFICATION OF CONDITIONS.**—Timber may be subjected to the following conditions:—

It may be kept constantly dry; at least, practically.
It may be constantly wet in fresh water.
It may be constantly damp.
It may be alternately wet and dry.
It may be constantly wet in sea water.

3. **TIMBER KEPT CONSTANTLY DRY** will last for centuries. The roof of Westminster Hall is more than 450 years old. In Stirling Castle are carvings in oak, well preserved, over 300 years old; and the trusses of the roof of

* *Hunt's Merchants' Magazine.*

† *New American Cyclopedia*, vol. xiii., p. 734.

the Basilica of St. Paul, Rome, were sound and good after 1,000 years of service.* The timber dome of St. Mark, at Venice, was in good condition 850 years after it was built.†

Artificial preservatives seem to be unnecessary under this condition.

4. TIMBER KEPT CONSTANTLY WET IN FRESH WATER, under such conditions as to exclude the air, is also very durable. The pillars upon which dwellings of the Canaries rest were put in their present place in 1402, and they remain sound to the present time.‡ The utensils of the lake dwellings of Switzerland are supposed to be at least 2,000 years old.§

The piles of the old London Bridge were sound 800 years after they were driven. The piles of bridge built by Trajan, after having been driven more than 1,600 years, were found to be petrified four inches, the rest of the wood being in its ordinary condition.§

Beneath the foundation of Savoy Place, London, oak, elm, beech, and chestnut piles and planks were found in a perfect state of preservation after having been there 650 years.§

While removing the old walls of Tunbridge Castle, Kent, there was found, in the middle of a thick stone wall, a timber curb which had been enclosed for 700 years.§

It is doubtful if artificial preparations would have prolonged the life of the timber in these cases.

5. TIMBER IN DAMP SITUATIONS.—Timber, in its native state, under these circumstances, is liable to decay rapidly from the disease called "dry rot." In dry rot the germs of the fungi are easily carried in all directions in a structure where it has made its appearance, without actual contact between the sound and decayed wood being necessary; whereas the communication of the disease resulting from wet rot takes place only by actual contact. The fungus is not the cause of the decay, but only converts corrupt matter into new forms of life.‖

There are three conditions which are at our command for prolonging the life of timber in damp situations:—

1st. Thoroughly season it;

2d. Keep a constant circulation of air about it; and

3d. Cover it with paint, varnish, or pitch.

The first condition is essential, and may be combined with either or both of the others.

By seasoning we do not mean simply drying so as to expel the water of the sap, but also a removal or change of the albuminous substances. These are fermentable substances, and when both are present they are ever ready, under suitable circumstances, to promote decay. The cellulose matter of the woody fibre is very durable when not acted upon by fermentation, and it is this that we desire especially to protect.

---

* The *London Builder*, vol. ii., p. 616.

† *Modern Carpentry*, Silloway, p. 40.

‡ *Journal of the Frank. Inst.*, 1870.

§ *Modern Carpentry*, Silloway, p. 39.

‖ "There is no reason to believe that fungi can make use of organic compounds in any other than a state of decomposition."—Carpenter's *Comp. Physiology*, p. 165. (See also *Encyclopædia Britanica* on this subject.)

Unseasoned timber which is surrounded by a dead air decays very rapidly. The timber of many modern constructions is translated from the forests and enclosed in a finished building in a few weeks, and unless it is subject to a free circulation of air it inevitably decays rapidly.*

Thorough ventilation is indispensable to the preservation of even well-seasoned *naked* wood in damp localities. The rapid decomposition of sills, sleepers, and lower floors is not surprising where neither wall-gratings nor ventilating flues carry off the moisture rising from the earth, or foul gases evolved in the decay of the surface mould. In the close air of cellars, and beneath buildings, the experiments of Pasteur detected the largest percentage of fungi spores. Remove the earth to the foot of the foundation, and fill in the cavity with dry sand, plaster-rubbish, etc., or lay down a thick stratum of cement to exclude the water, and provide for a complete circulation of air, and lower floors will last nearly as long as upper ones.†

A covering of paint, pitch, varnish, or other impervious substance upon undried timber is very detrimental, for by it all the elements of decay are retained and compelled to do their destroying work. The folly of oiling, painting, or charring the surface of unseasoned wood is therefore evident. Owing to this blunder alone, it is no unusual thing to find the painted wood-work of older buildings completely rotted away, while the contiguous naked parts are perfectly sound.

While an external application of coal tar promotes the preservation of dry timber, nothing can more rapidly hasten decay than such a coating upon the surface of green wood. But this mistake is often made, and dry rot does the work of destruction.‡ Carbonizing the surface also increases the durability of dry, but promotes the decay of wet timber. Farmers very often resort to one of the latter methods for the preservation of their fence-posts. Unless they discriminate between green and seasoned timber, these operations will prove injurious instead of beneficial.

There are numerous methods for promoting the process of seasoning. Some have in view simply *drying*, a process which is important in itself, but which will not in itself prevent decay in damp situations unless the moisture be permanently excluded. Some dry with hot air, and some with steam. In the latter case, if the steam be superheated the process is very rapid, but it seems to damage the life of the timber.

Others have in view the expulsion of the albuminous substance. Water-soaking the logs and afterwards drying the lumber, seems to be a cheap and quite effectual mode. But there are many patented processes for securing this end, or for changing the albuminous substances; and in many cases the latter end is not only secured, but the salts which are used act directly upon the cellulose and lignite of the wood, thereby greatly promoting its durability.

---

* For an account of the rapid destruction of the floors and joists of the Church of the Holy Trinity, Cork, Ireland, by dry rot, see *Civil Engineer's Journal*, vol. xii., p. 303. For an account of the decay of floors, studs, &c., in a dwelling, see the *London Builder*, vol. vi., p. 34.

"In some of the mines in France the props seldom last more than fifteen months."—*Annales des Mines.*

† *The Builder*, vol. xi., page 46.

‡ According to Col. Berrien, the Michigan Central Railroad bridge, at Niles, was painted, *before seasoning*, with "Ohio fire-proof paint," forming a glazed surface. About five years after, it was so badly dry-rotted as to require rebuilding.

The following are the principal processes which have been used:—Mr. P. W. Barlow's patent* provided for exhausting the air from one end of the log, while one or more atmospheres press upon the other end. This artificial aërial circulation through the wood is prolonged at pleasure. However excellent in theory, this process is not practicable.

By another method, the smoke and hot gases of a coal fire are conveyed among the lumber, placed in a strong draft. Some writers recommend the removal of the bark one season before felling the tree. All good authorities agree *that the cutting should take place in the winter season.*†

Kyan's process, which consists in the use of corrosive sublimate, was patented in 1832. His specific solution ‡ was one pound of chloride of mercury to four gallons of water. Long immersion in the liquid in open vats, or great pressure upon both solution and wood, in large wrought-iron tanks, is necessary for the complete injection of the liquid. The durability of well kyanized timber has been proved, but the expensiveness of the operation will long forbid its extensive adoption.

For "Burnettizing," § a solution of chloride of zinc—one pound of salt to ten gallons of water—is forced into the wood under a pressure of 150 lbs. per square inch.

Boucherie employs a solution ‖ of sulphate of copper one pound to water twelve and a half gallons, or pyrolignite of iron one gallon to six gallons of water. He enclosed one end of the green stick in a close-fitting collar, to which is attached an impervious bag communicating through a flexible tube with an elevated reservoir containing the salt liquid. Hydrostatic pressure soon expels the sap at the opposite end of the log. When the solution makes its appearance also the process is completed.

He finds the fluid will pass *along* the grain, a distance of 12 feet, under a lower pressure than is required to force it *across* the grain, three-fourths of an inch. The operation is performed upon green timber with the greatest facility.¶

In 1846, eighty thousand sleepers of the most perishable woods, impregnated, by Boucherie's process, with sulphate of copper, were laid down on French railways. After nine years' exposure, they were found as perfect as when laid.** This experiment was so satisfactory that most of the railways of that empire at once adopted the system. We would suggest washing out the sap with water, which would not coagulate its albumen. The solution would appropriately follow.

Both of the last-named processes are comparatively cheap. The manufacturing companies of Lowell, Massachusetts, have an establishment for "Burnettizing" timber,†† in which they prepare sticks fifty feet in length. Under

---

* *Civ. Eng. Jour.*, vol. xix., p. 422.

† Experiments detailed in the *Cosmos* show conclusively that winter-cut pine is stronger and more durable than that cut at any other season of the year.—*Ann. Sc. Discovery* for 1861, p. 346.

"Oak trees felled in the winter make the best timber."—*The Builder*, 1859, page 138.

‡ *Civ. Eng. Jour.*, vol. v., page 202.

§ *Civ. Eng. Jour.*, vol. xiv., p. 471. Invented by Burnett in 1838.

‖ *Civ. Eng. Jour.*, vol. xx., p. 405.

¶ As a modification of this method he also cut a channel in the wood throughout the circumference of the tree, fitted a reservoir thereunto, and poured in the liquid. The vital forces speedily disseminated the solution throughout the tree.

** *Jour. of the Frank. Inst.*, vol. xxxii., pp. 2, 3.

†† *New American Cyclopædia.*

a pressure of 125 pounds per square inch they inject from two to eight ounces of the salt into each cubic foot of wood. The cost, in 1861, was from $5 to $6 per 1,000 feet, board measure.* Boucherie's method must be still cheaper. It costs less than creosoting by one shilling per sleeper.†

An American engineer, Mr. Hewson, for injecting railroad sleepers, proposes a vat deep enough for the timbers to stand in upright. The pressure of the surrounding solution upon the lower ends of the sticks will, he thinks, force the air out at their upper extremities, kept just above the surface of the solution, after which the latter will rise and impregnate the wood. In 1859 he estimated chloride of zinc at 9 cents per pound, sulphate of copper at 14 cents per pound, and pyrolignite of iron at 23 cents per gallon. He found the cost of impregnating a railway tie with sufficient of those salts to prevent decay, to be: for the chloride of zinc 2·8 cents, for blue vitriol 3·24 cents, for pyrolignite of iron 7·5 cents.‡

By Earle's process the timber is boiled in a solution of one part of sulphate of copper, three parts of the sulphate of iron, and one gallon of water to every pound of the salts. A hole was bored the whole length of the piece before it was boiled. It was boiled from two to four hours, and allowed to cool in the mixture.

Ringold and Earle invented the following process:—A hole was made the whole length of the piece, from one-half to two inches in diameter, and boiled from two to four hours in lime-water. After the piece was dried the hole was filled with lime and coal tar. Neither of these methods was very successful.

A Mr. Darwin suggests that the piece be soaked in lime-water, and afterwards in sulphuric acid, so as to form gypsum in the pores.

Bethell's process consists in forcing dead oil into the timber. This is called creosoting.§ He inclosed the timber and dead oil in huge iron tanks, and subjected them to a pressure varying between 100 and 200 pounds per square inch, at a temperature of 120° F. about twelve hours. From eight to twelve pounds of oil are thus injected into each cubic foot of wood. Lumber thus prepared is not affected by exposure to air and water, and requires no painting.‖ A large number of English railway companies have already adopted the system.¶ Eight pounds of oil per cubic foot is sufficient for railway sleepers.**

One writer has said that if creosote has ever failed to prevent decay, it has been because of an improper treatment, or because the oil was deficient in carbolic acid.

Sir Robert Smirke was one of the first architects to use this process, and when examined before a *Committee on Timber*, stated that this process does not

---

* The Philadelphia, Wilmington and Baltimore Railroad Company have used the process since 1860 with complete success. The Union Pacific Railroad Company have recently erected a large building for this purpose. Their cylinder is 75 feet long, 61 inches in diameter, and capable of holding 250 ties. They "Burnettize" two batches per day.—*Report on Pacific Railroad*, by Col. Simpson, 1865.

† *Jour. Frank. Inst.*, vol. xxxii., pp. 2, 3.

‡ Ibid., vol. xxvii., p. 8.

§ "Creosote from coal undoubtedly contains two homologous bodies, $C_{12}H_6O_2$ and $C_{14}H_8O_2$, the first being carbolic and the second crysilic acid."—Ure's *Dict. of Arts, Manu., and Mines*, vol. ii., p. 623.

‖ Ure's *Dict. of Manu. and Mines.*

¶ The Great Western, North-Eastern, Bristol and Exeter, Stockton and Darlington, Manchester and Birmingham, aad London and Birmingham.—Ure's *Dict. of Manu. and Mines.*

** *Jour. Frank. Inst.*, vol. xliv., p. 275.

diminish the strength of the material which is operated upon. He afterwards said, "I cannot rot creosoted timber, and I have put it to the severest test I could apply."

The odor of creosote makes it objectionable for residences and public buildings.

Mr. S. Beer, of New York City, invented a mode of preserving timber by boiling it in borax with water. But this process has been objected to on the ground that it is not a good protection against moisture.

Common salt is known to be a good preservative in many cases. According to Mr. Bates's opinion,* it answers a good purpose in many cases if the pieces to which it is exposed are not too large.

6. TIMBER ALTERNATELY WET AND DRY.—The surface of all timber exposed to alternations of wetness and dryness, gradually wastes away, becoming dark-colored or black. This is really a slow combustion, but is commonly called wet rot, or simply rot. Other conditions being the same, the most dense and resinous woods longest resist decomposition. Hence the superior durability of the heart-wood, in which the pores have been partly filled with lignine, over the open sap-wood, and of dense oak and lignum-vitæ over light poplar and willow. Hence, too, the longer preservation of the pitch-pine and resinous "jarrah" of the East, as compared with non-resinous beech and ash.

Density and resinousness exclude water. Therefore our preservatives should increase those qualities in the timber. Fixed oils fill up the pores and increase the density. Staves from oil-barrels and timbers from whaling ships are very durable. The essential oils resinify, and furnish an impermeable coating. But pitch or dead oil possesses advantages over all known substances for the protection of wood against changes of humidity. According to Professor Letheby,† dead oil, 1st, coagulates albuminous substances; 2d, absorbs and appropriates the oxygen in the pores, and so protects from eremacausis; 3d, resinifies in the pores of the wood, and thus shuts out both air and moisture; and 4th, acts as a poison to lower forms of animal and vegetable life, and so protects the wood from all parasites. All these properties specially fit it for impregnating timber exposed to alternations of wet and dry states, as, indeed, some of them do, for situations damp and situations constantly wet. Dead oil is distilled from coal-tar, of which it contains about .30, and boils between 390° and 470° Fahr. Its antiseptic quality resides in the creosote it contains. One of the components of the latter, carbolic acid (phenic acid, phenol), $C_{12}H_6O_2$, the most powerful antiseptic known, is able at once to arrest the decay of every kind of organic matter.‡ Prof. Letheby estimates this acid at ½

---

* Report of the Commissioner of Agriculture.

† *Civ. Eng. Jour.*, vol. xxiii., p. 216.

‡ "I have ascertained that adding one part of the carbolic acid to five thousand parts of a strong solution of glue will keep it perfectly sweet for at least two years. . . . . Hides and skins, immersed in a solution of one part of carbolic acid to fifty parts of water, for twenty-four hours, dry in air and remain quite sweet."—Prof. Crace Calvert, *Ann. Sc. Discov.*, 1865, p. 55.

"Carbolic acid is sufficiently soluble in water for the solution to possess the power of arresting or preventing spontaneous fermentation. Saturated solutions act on animals and plants as a virulent poison, though containing only five per cent. of the acid."—*Civ. Eng. Jour.*, vol. xxii., p. 216.

"Parasites and other worms are instantly killed by a solution containing only one-half per cent. of acid, or by exposure to the air containing a small portion of the acid. . . . . By examining the action on a leaf, we find the albumen is coagulated. All animals with a naked skin, and those that live in water, die sooner than those that live in air and have a solid envelope."—Dr. I. Lemaire, *Ann. Sc. Discov.*, 1855, p. 238.

to 6 per cent. of the oil. Chrysilic acid $C_{14}H_8O_2$, the homologue of carbolic acid, and the other component of creosote, is not known to possess preservative properties.

Creosoting, or Bethell's process, is the most valuable of all the well-tried processes in this case. For railway sleepers eight pounds of oil per cubic foot of timber is sufficient.* If the timber is dry, a coating of coal-tar, paint, or resinous substance, is valuable.

A Mr. Heinmann, of New York City, proposes the following process, which appears to be very promising :—

The sap is first expelled and then the timber is injected with common rosin. The latter is introduced while in a liquid state, under high pressure, while in vessels especially constructed for the purpose.

In an experiment made by Prof. Ogden, one cubic foot of green wood absorbed 8.96 pounds of rosin, while a cubic foot of well-seasoned wood absorbed only 2.66 pounds. The strength of the timber was increased by this process, as is shown by the following experiment :—

Wood treated with Rosin.			Wood in its Natural State.		
Breaking Weight. Pounds.	Quality.	Grain.	Breaking Weight. Pounds.	Quality.	Grain.
163.5	Checked.	Straight.	98.5	Sound.	Slant.
193	Sound.	"	103.0	"	"
171.5	"	"	116.0	"	Straight.
72.5	Checked.	Cross.	57.5	"	"
57.5	"	Slant.	46.00	"	"
57.5	"	"	46.0	"	"
121.0	"	"	71.0	"	"
155.5	"	Cross.	84.0	"	"

It is found by experiment that wood thus treated is not as flammable as air-dried wood. This is accounted for from the fact that a kind of inflammable slag is deposited over the surface immediately after the rosin begins to burn.

The chief advantages which are claimed for this method are more theoretical than practical, as it has not yet had sufficient time to test its practical merits, and it may, like many other processes, disappoint the hopes of its strongest advocates and well-wishers.

7. TIMBER CONSTANTLY WET IN SALT WATER.—We have not to guard against decay when timber is in this situation. Teredo navalis, a mollusk of the family Tubicolària, Lam , soon reduces to ruins any unprotected submarine construction of common woods. We quote from a paper read before the "Institute of Civil Engineers," England, illustrating the ravages of this animal :—

"The sheeting at Southend pier extended from the mud to eight feet above low-water mark. The worm destroyed the timber from two feet below the

* *Jour. Frank. Inst.*, vol. xliv., p. 275.

surface of the mud to eight feet above low-water mark, spring-tide; and out of 38 fir-timber piles and various oak-timber piles, not one remained perfect after being up only three years." * Specimens of wood, taken from a vessel that had made a voyage to Africa, are in the museum, and show how this rapid destruction is effected.

"None of our native timbers are exempt from these inroads. Robert Stephenson, at Bell Rock, between 1814 and 1843,† found that green heart oak, beef-wood, and bullet-tree were not perforated, and teak but slightly so. Later experiments show that the "jarrah" of the East, also, is not attacked.‡ The cost of these woods obliges us to resort to artificial protection.

"The teredo never *perforates* below the surface of the sea-bottom, and probably does little injury above low-water mark. Its minute orifice, bored across the grain of the timber, enlarges inwards to the size of the finger, and soon becomes parallel to the fibre. The smooth circular perforation is lined throughout with a thin shell, which is sometimes the only material separating the adjacent cells. The borings undoubtedly constitute the animal's food, portions of woody fibre having been found in its body.§ While upon the surface only the projecting siphuncles indicate the presence of the teredo, the wood within may be absolutely honey-combed with tubes from one to four inches in length.

"It was naturally supposed that poisoning the timber would poison or drive away the teredo, but Kyan's, and all other processes employing solutions of the salts of metals of alkaline earths, signally failed. This, however, is not surprising. The constant motion of sea-water soon dilutes and washes away the small quantity of soluble poison with which the wood has been injected. If any albuminate of a metallic base still remains in the wood, the poisonous properties of the injection have been destroyed by the combination. Moreover, the lower vertebrates are unaffected by poisons which kill the mammals. Indeed, it is now known that certain of the lower forms of animal life live and even fatten on such deadly agents as arsenic.‖

"Coatings of paint or pitch are too rapidly worn away by marine action to be of much use, but timber, thoroughly creosoted with ten pounds of dead oil per cubic foot, is perfectly protected against teredo navalis. All recent authorities agree upon this point. In one instance, well authenticated, the mollusk reached the impregnated heart-wood by a hole carelessly made through the injected exterior. The animal pierced the heart-wood in several directions, but turned aside from the creosoted zone.¶ The process and cost of "creosoting" have already been discussed."

A second destroyer of submarine wooden constructions is limnoria terebrans, (or L. perforata, Leach) a mollusk of the family Assellotes, Leach, resembling the sow-bug. It pierces the hardest woods with cylindrical, perfectly smooth, winding holes, $\frac{1}{20}$th to $\frac{1}{15}$th of an inch in diameter, and about two inches deep.** From ligneous matter having been found in its viscera, some have concluded that the limnora feeds on the wood, but since other mollusks of the same ge-

* *Civ. Eng. Jour.*, vol. xii., p. 382.
† *The Builder* for 1862, p. 511.
‡ *Civ. Eng. Jour.*, vol. xx., p. 17.
§ *Civ. Eng. Jour.*, vol. xii., p. 382. Also *Dict. Univ. d'Hist. Natur.* tome xii.
‖ *British and Foreign Medical Review.*
¶ *Civ. Eng. Jour.*, vol. xii., p. 191.
** *Dict. Univ. d'Hist. Natur.*

nus, Pholas, bore and destroy stone-work, the perforation may serve only for the animal's dwelling. The lumnoria seems to prefer tender woods, but the hardest do not escape. Green-heart oak is the only known wood which is not speedily destroyed.* At the harbor of Lowestoft, England, square fourteen-inch piles were, in three years, eaten down to four inches square.†

While all agree that no preparation, if we except dead oil, has repelled the limnoria, an eminent engineer has cited three cases in which that agent afforded no protection.‡

We do not find that timber impregnated with water-glass has been tested against this subtle foe. The experiment is certainly worthy of a trial.

A mechanical protection is found in thickly studding the surface of the timber with broad-headed iron nails. This method has proved successful. Oxydation rapidly fills the interstices between the heads, and the outside of the timber becomes coated with an impenetrable crust, so that the presence of the nails is hardly necessary.

In conclusion, we cannot but express surprise that so little is known in this country concerning preservative processes. Their employment seems to excite very little interest, and the very few works where they are being tested attract hardly any attention. Those railroads which have suspended their use assign no reasons, and those upon which the timber is injected publish no reports concerning the advantages of their particular methods. Even the National Works, upon which Kyan's process was formerly employed, have laid it aside, and now subject lumber to dampness and alternations of wetness and dryness, without any preparation beyond seasoning. When sleepers cost fifty cents and creosoting thirty cents each, it is cheaper to hire money at seven per cent., compound interest, than to lay new sleepers at the end of seven years. Allowing any ordinary price for the removal of the old and laying down the new ties, the advantage of using Bethell's process seems evident. If some cheaper method will produce the same effects, the folly of neglecting *all* means which aim at increasing the durability of the material is still more palpable.

---

* *Civ. Eng. Jour.*, vol. xxv., p. 206.
† Ibid., vol. xvi., p. 76.
‡ Ibid., vol. xxv., p. 206.

8. THE FOLLOWING IS A SUMMARY of the different processes that have been invented from time to time, with the names of their inventors:—

Names of Inventors.	Chemicals Used.	Manner of using them.
Bethell	Creosote, or pitch-oil	By injection.
Kyan	Chloride of mercury	" "
Margery	Sulphate of copper	" "
Burnett	Chloride of zinc	" "
Ransome	Liquid silicate of potassa	" "
LeGras	Manganese, lime, and creosote	" "
Margary	Solution of acetate of copper	" "
Payne	Sulphate of iron, carbonate of soda	" "
Bouchere	Pyrolignate of iron, sulphate of copper	" "
Gemini	Tar	" "
Heinmann	Rosin, or colophony	" "
Earle	Sulphate of copper	" boiling.
Ringold	Lime	" "
Tregold	Sulphate of iron	" "
S. Beer	Borax	" "
Dorset and Blythe	Same as Bouchere	By means of a vacuum and injection.
Huting and Boutigny	Oil of schist, tar, pitch, and shellac	By immersion, and by fire or burning.
Vernet	Arsenic	By saturation.
	Salt	External application.

APPENDIX II.

TABLE

Of the Mechanical Properties of the Materials of Construction.

NOTE.—The capitals affixed to the numbers in this table refer to the following authorities :—

B. Barlow. Report of the Commissioners of the Navy, etc.
Be. Bevan.
Bn. Buchanan.
Br. Belidor, Arch. Hydr.
Bru. Brunel.
C. Couch.
Cl. Clark.
D. Darcel, Annales for 1858.
D. W. Daniell and Wheatstone. Report on the stone for the Houses of Parliament.
E. Eads.
F. Fairbairn.
G. Grant.
H. Hodgkinson. Report to the British Association of Science, etc.
Ha. Haswell. Eng. and Mech. Pocket-Book, 1869.
J. Journal of Franklin Institute, vol. XIX., p. 451.
K. Kirwan.
Ki. Kirkeldy.
La. Lamé.
M. Mischembroeck. Introd. ad. Phil. Nat. I.
Ma. Mallet.
Mi. Mitis.
Mt. Mushet.
Pa. Colonel Pasley.
R. Roudelet. L'Art de Batir, IV.
Ro. Roebling.
Re. Reunie. Phila. Trans., etc.
S. Styffe. On Iron and Steel.
T. Thompson.
Te. Telford.
Tr. Tredgold. Essay on the Strength of Cast Iron.
W. Watson.
Wa. Major Wade.
Wn. Wilkinson.

* Calculated from the experiments of Fairbairn and Hodgkinson.

NAMES OF MATERIALS.	Weight of one cubic ft. in lbs. δ	Tenacity per sq. inch in lbs. T.	Crushing Force per square inch in lbs. C.	Modulus of Rupture. R.	Coefficient of Elasticity. E.
METALS.					
Antimony—					
Cast	281·25	1,066 M.			
Bismuth	613·87	3,250 M.			
Brass—					
Cast	525·00	17,968 Re.	10,304 Re.	....	9,170,000
Wire-drawn	534·00	....	....	....	14,230,000
Copper—					
Cast	537·93	19,072	29,272 Re.		
Sheet	549·06	32,184			
Wire-drawn	560·00	61,228			
In Bolts	....	48,000			
Iron.					
Cast Iron.					
Old Park	....	....	....	48,240 T.	18,014,400 T.
Carron, No. 2—					
Cold Blast	441·62	16,683 H.	106,375 H.	38,556 H.	17,270,500 H.*
Hot Blast	440·37	13,505 H.	108,540 H.	37,503 H.	16,085,000 H.*
Carron, No. 3—					
Cold Blast	443·37	14,200 H.	115,442 H.	35,980 F.*	16,246,966 F.
Hot Blast	441·00	17,775 H.	133,440 H.	42,687 F.*	17,873,100 F.

TABLE—*Continued.*

NAMES OF MATERIALS.	Weight of one cubic ft. in lbs. δ	Tenacity per sq. inch in lbs. T.	Crushing Force per square inch in lbs. C.	Modulus of Rupture. R.	Coefficient of Elasticity. E.
Iron.					
Cast Iron.					
Devon, No. 3—					
Cold Blast	455·93	....	....	36,288 H.*	22,907,700 H.
Hot Blast	451·81	29,107 H.	145,435 H.	43,497 H.*	22,473,650 H.
Buffery, No. 1—					
Cold Blast	442·43	17,466 H.	93,366 H.	37,503 H.*	15,381,200 H.
Hot Blast	437·37	13,434 H.	86,397 H.	35,316 H.*	13,730,500 H.
Coed Talon, No. 2—					
Cold Blast	434·06	18,855 H.	81,770 H.	33,453 F.*	14,313,500 F.
Hot Blast	435·50	16,676 H.	82,739 H.	33,696 H.*	14,322,500 F.
Elsicar, No. 1—					
Cold Blast	439·37	....	....	34,587 F.*	13,981,000 F.
Milton, No. 1—					
Hot Blast	436·00	....	....	29,889 F.*	11,974,500 F.
Muirkirk, No. 1—					
Cold Blast	444·56	....	....	36,693 F.*	14,003,550 F.
Hot Blast	434·56	....	....	33,850 F.*	13,294,490 F.
Morris Stirling's 2d quality (See also p. 52.)	....	25,764	119,000		
Gun Metal—					
American	....	....	14,000 to 34,000 Wa.	....	27,548,000 Wa.
Extra Specimens	595·00	....	45,970 Wa.		
Steel.					
Hammered Cast Steel, from F. Krupp	....	91,000 / 122,000 S.	....	....	31,359,000 S.
Tempered	....	171,000 S.			
Bessemer Steel, from Högbo, marked 10	....	140,945 S.	....	....	31,819,000 S.
Bessemer Steel, Eng. Mean of four experiments	485·37	88,415 F.	225,568 F.	....	29,215,000 F.
Naylor, Vickers & Co. Crucible Steel	488·70	108,099 F.	225,568 F.	....	30,278,000 F.
Mushet's Steel—					
Soft	492·50	93,616 F.	....	....	31,901,000 F.
Cast Steel—					
Soft	486·25	120,000			
Not Hardened	....	....	198,944 Wa.		
Mean Temper	....	....	381,985 Wa.		
Razor Tempered	490·00	150,000	....	....	29,000,000
Steel Wire Rope—					
Fine Wire	....	40,000 Ro.			
Cast Steel	....	242,100 See page 25.			
Wrought Iron.					
English	481·20	57,300 La.	See page 53.		From 22,000,000 to 28,000,000
In Bars	475·50 / 487·00	57,300 La.			
Hammered	....	67,200 Bru.			
Russian	....	60,480 La.			
Swedish, in bars	....	71,680 R.			
English, in wire 1-10 inch diam.	....	80,000 Te. / 96,000 Te.			
Russian, in wire; diam. 1-20 to 1-30 inch	....	134,000 La. / 203,000 La.			

TABLE—*Continued.*

NAMES OF MATERIALS.	Weight of one cubic ft. in lbs. δ	Tenacity per sq. inch in lbs. T.	Crushing Force per square inch in lbs. C.	Modulus of Rupture. R.	Coefficient of Elasticity. E.
Wrought Iron.					
Rolled in sheets and cut crosswise	....	40,320 Mi.			From 22,000,000 to 28,000,000.
Cut lengthwise	....	31,360 Mi.			
In chains, oval links, iron ⅞ in. diam	....	48,160 Br.			
Wire, American	....	73,600 Ha.			
Lake Superior and Iron Mountain Charcoal Bloom	....	90,000 Ha.			
Missouri Iron, bar	....	47,909 J.			
Tennessee, bar, 21 exp	....	52,099 J.			
Salisbury, Ct., 40 exp	....	58,009 J.			
Centre Co. Pa., 15 exp	....	58,400 J.			
Phillipsburgh Wire, Pa. Diam. 0·333 in., 13 exp	....	84,186 J.			
Diam. 0.190 in., 5 exp	....	73,888 J.			
Diam. 0.156 in., 5 exp	....	89,162 J.			
Mean of 188 rolled bars	....	57,557 Ki.			
Mean of 167 plates lengthwise	....	50,737 Ki.			
Mean of 160 plates crosswise	....	46,171 Ki.			
Low Moor, bars	....	60,364 Ki.			
Swedish, forged	....	41,000 Ki. 50,000 Ki.			
Hammered Bessemer Iron, from Högbo	....	....	....	....	32,320,000 S.
Low Moor Rolled Puddled Iron	....	....	....	....	21,976,000 S.
Rolled Iron, Swedish, charcoal heath	....	65,000 S.	....	....	27,000,000 S.
Lead, cast, English	717·45	1,824 Re.			
Lead Wire	705·12	2,581 M.			
Silver, standard	644·50	40,902 M.			
Tin, cast	455·68	5,322 M.	....	....	4,608,000 Tr.
Zinc	439·25	....	....	....	13,680,000 Tr.
STONE—NATURAL AND ARTIFICIAL.					
Granites.					
Aberdeen, blue	164	....	10,914 Re.		
Cornish	166	....	6,356 Re.		
Killiney, very felspathic	....	....	10,780 Wn.		
Mount Sorrell, granite	166	....	12,286 F.		
Sandstones.					
Caithness Pavement	....	....	6,493 Bn.		
Dundee Sandstone	158	....	6,630 Re.		
Derby Grit, a red, friable Sandstone	148	....	3,142 Re.		
Do. from another quarry	156	....	4,345 Re.		
Limestones.					
Limestone, Magnesian (Grafton, Ill.)	....	....	17,000 E.	....	Same as W't. Iron. E.

TABLE—*Continued.*

NAMES OF MATERIALS.	Weight of one cubic ft. in lbs. δ	Tenacity per sq. inch in lbs. T.	Crushing Force per square inch in lbs. C.	Modulus of Rupture. R.	Coefficient of Elasticity. E.
Limestones.					
Limestone, compact.........	162	....	7,713 Re.		
Limestone, Kerry, Listowel Quarry, Eng.............	....	....	18,043 Wn.		
Chalk	....	....	501 Re.		
Other Stones.					
Alabaster (Oriental), white..	170				
Marble, statuary...........	....	....	3,216 Re.		
Do. white Italian, veined.	165	....	9,681 Re.	1,062	25,200,000 T.
Do. black Galloway......	168	....	9,219 Re.	2,664	
Portland Stone (Oolite)......	151	....	3,792 Re.		
Valentia, Kerry (slate stone)..	....	....	10,943 Wn.		
Green Stone, from Giant's Causeway................	....	....	17,220 Wn.		
Quartz Rock, Holyhead (across lamination)........... ...	....	....	25,500 Ma.		
Quartz Rock (parallel to lamination)..................	....	....	14,000 Ma.		
Gravel.....................	120				
Green Moor................	158	....	2,010 Re.		
Artificial Stone.					
Brick, red.................	135·5	280	808 Re.		
Brick, pale red.............	130·31	300	562 Re.		
Brick, common.............	....	....	800 to 4,000 Ha.		
Bire Brick, Stourbridge......	....	....	1,717 Re.		
Brick, Stock...............	....	....	2,177 Ha.		
Bricks set in cement (bricks not very hard)...........	....	....	521 Cl.		
Brick Masonry, common.....	....	....	500 to 800 Ha.		
Cement, Portland, with sand.	....	92 to 284 D.			
Cement, Portland, with no sand......................	....	427 to 711			
Cement, Portland...........	....	....	1,000 to 5,900 G.		
Chalk	116·81	....	334 Re.		
Glass, Plate................	153·31	9,420	....		
Mortar.....................	107	50	120 to 240 Ha.		
TIMBER.					
Acacia, English.............	47·37	16,000 Be.	....	11,202 B.	1,152,000 B.
Alder......................	50·00	14,186 M.	6,859 H.		
Apple Tree.................	49·56	19,500 Be.			
Ash { Ordinary state........	43·12	17,207 B.	8,683 H.	12,156 B.	1,644,800 B.
Ash { Very dry............	55·81		9,363 H.		
Bay Tree.	51·37	12,396	7,158 H.		
Bean, Tonquin.............	67·51	....	....	20,886 B.	2,601,600 B.
Beech { Ordinary...........	53·37	15,784 B.	7,733 H.	9,336 B.	1,353,600 B.
Beech { Very dry...........	45·12	17,850 B.	9,363 H.		

TABLE—*Continued.*

Names of Materials.	Weight of one cubic ft. in lbs. δ	Tenacity per sq. inch in lbs. T.	Crushing Force per square inch in lbs. C.	Modulus of Rupture. R.	Coefficient of Elasticity. E.
Timber.					
Birch, common	49·50	15,000	4,533 H. 6,402 H.	10,920 B.	1,562,400 B.
Birch, American	40·50	…………	11,663 H.	9,624 B.	1,257,600 B.
Box, dry	60·00	19,891 B.	10,299 H.		
Bullet Tree (Berbice)	64·31	…………	…………	15,636 B.	2,610,600 B.
Cane	25·00	6,300 Be.			
Cedar, Canadian	56·81	11,400 Be.	5,674 H.		
Crab Tree	47·80	…………	6,499 H.		
Deal—					
Christiana Middle	43·62	12,400	…………	9,864 B.	1,672,000 B.
Norway Spruce	21·25	17,600			
English	29·37	7,000			
Red	…………	…………	5,748 H.		
White	…………	…………	6,741 H.		
Elder	43·43	10,230	8,467 H.		
Elm, seasoned	36·75	13,489 M.	10,331 H.	6,078 B.	699,840 B.
Fir—					
New England	34·56	…………	…………	6,612 B.	2,191,200 B.
Riga	47·06	11,549 to 12,857 B.	5,748 to 6,586 H.	6,648 B. 7,572 B.	1,328,800 869,600 B.
Hazel	53·75	18,000 Be.			
Lance Wood	63·87	24,696			
Larch—					
Green	32·62	10,220 B.	3,201 H.	4,992 B.	897,600 B.
Dry	35·00	8,900 B.	5,568 H.	6,894 B.	1,052,800 B.
Lignum vitæ	76·25	11,800 M.			
Mahogany, Spanish	50·00	16,500	8,198 H.		
Maple, Norway	49·56	10,584			
Oak—					
English	58·37	17,300 M.	4,684 to 9,509 H.	10,032 B.	1,451,200 B.
Canadian	54·50	10,253	4,231 to 9,509 H.	10,596 B.	2,148,800 B.
Dantzic	47·24	12,780	…………	8,742 B.	1,191,200 B.
Adriatic	62·06	…………	…………	8,298 B.	974,400 B.
African Middle	60·75	…………	…………	13,566 B.	2,283,200 B.
Pear Tree	41·31	…………	7,518 H.		
Pine—					
Pitch	41·25	7,818 M.	…………	9,792	1,225,600 B.
Red	41·06	…………	5,375 H.	8,946 B.	1,840,000 B.
American Yellow	28·81	…………	5,445 H.	…………	1,600,000 Tr.
Plum Tree	49·06	11,351	3,657 to 9,367 H.		
Poplar	23·93	7,200	3,107 to 5,124 H.		
Teak, dry	41·06	15,000 B.	12,101 H.	14,722 B.	2,414,400 B.
Walnut	41·93	8,130 M.	6,635 H.	…………	306,000
Willow, dry	24·37	14,000 Be.			

ERRATA.

Page 14, line 9, for $P = 13, 934,000\frac{\lambda e}{l} = 2,907,432,000\frac{\lambda e}{l^2}$, read,

$$P = 13,934,000\frac{\lambda e}{l} - 2,907,432,000\frac{\lambda^2}{l^2}e.$$

" 22, " 5, for d^2t, read, dt^2.

" " " at the middle of the page, for *Resiliance of Prisms*, read, *Resilence of Prisms.*

" " " 10 from the bottom, for $\frac{2EK}{l}\lambda^2$, read, $\frac{EK}{l}\lambda^2$.

" 26, " 3, for exponetials, read, exponentials.

" 36, " 10, for rods *or* rivet iron, read, rods *of* rivet iron.

" 53, " 2, for No. 1, read, No. 2.

" 56, " 3 from the bottom, for x, read, y.

" 68, " at the bottom of the table, for Mean 6852, read, 685·2.

" 72, " 15, for equation (26), read, equation (24).

" 96, " 1, for **82,** read, **86.**

www.ingramcontent.com/pod-product-compliance
Lightning Source LLC
LaVergne TN
LVHW010249110826
845151LV00004B/1423

* 9 7 8 1 4 2 5 5 2 2 6 2 9 *